Generic Multi-Agent Reinforcement Learning Approach for Flexible Job-Shop Scheduling

Schirin Bär

Generic Multi-Agent Reinforcement Learning Approach for Flexible Job-Shop Scheduling

Schirin Bär
Nürnberg, Germany

D 82 (Diss. RWTH Aachen University, 2021)

ISBN 978-3-658-39178-2   ISBN 978-3-658-39179-9   (eBook)
https://doi.org/10.1007/978-3-658-39179-9

Responsible Editor: Stefanie Eggert
This Springer Vieweg imprint is published by the registered company Springer Fachmedien
Wiesbaden GmbH, part of Springer Nature.
The registered company address is: Abraham-Lincoln-Str. 46, 65189 Wiesbaden, Germany

Danksagung

Die Zeit, in der diese Arbeit entstand, war prägnet für mich und wurde geprägt durch intelligente und inspirierende Menschen an meiner Seite.

Ich bin unheimlich dankbar, Prof. Dr. Ing. Tobias Meisen als meinen betreuenden Professor gewonnen zu haben. Jede Besprechung mit ihm war äußerst wertstiftend und durch seine inspirierende, energetische und positive Art hat er mich dazu gebracht mich nach jedem unserer Meetings wieder motiviert an die Arbeit zu setzen und neue Experimente zu starten. Ich bin immer wieder beeindruckt davon, wie schnell er sich in neue Ideen eindenken und sie weiterentwickeln kann. Rückblickend waren diese Zutaten das Geheimrezept, um die Industriepromotion in kurzer Zeit erfolgreich zu meistern.

Auch Prof. Dr. phil. Ingrid Isenhardt möchte ich für die Betreuung der Arbeit meinen Dank aussprechen.

Meinem Lieblingskollegen, Co-Autor und Ehemann, Felix, bin ich unendlich arg dankbar für die Zeit und vor allem die Geduld, die er sich genommen hat, mich zu unterstützen, mich inhaltlich voranzubringen und vor allem mich motiviert zu halten. Mit ihm an meiner Seite fühlte sich der Weg so viel leichter an. Auch örtlich hat er mich begleitet, von Nürnberg über Mannheim nach Gera, bis er mich letztlich überzeugt hat den nächsten Lebensabschnitt mit ihm in Suzhou, China zu verbringen.

Mein Research-Team bei Siemens möchte ich besonders hervorheben, weil es einen maßgeblichen Teil zu unserer gemeinsamen Forschung, Veröffentlichungen und Präsentationen beigetragen hat, vor allem Sebastian, Danielle und Punit. Die Arbeit mit ihnen hat mir wahnsinnig viel Freude bereitet.

Ein großer Dank geht an meine weiteren Siemens Kollegen und Kolleginnen, mit denen die Arbeit wie ein Hobby war. Neben dem Spaß bei der Arbeit haben mich die vielen guten Gespräche und Workshops wahnsinnig weitergebracht. Vielen Dank an Jörn, Hans-Henning, Jupiter, Hans-Georg, Benjamin, André und vielen weiteren.

Auch meinen Kollegen und Kolleginnen an der RWTH Aachen und Universität Wuppertal möchte ich einen großen Dank aussprechen. Der Austausch war sehr motivierend und hilfreich. Vielen Dank vor allem an Vladimir, Daniel, André und allen Kollegen und Kolleginnen, die mich bei der Vorbereitung meiner Doktorprüfung unterstützt haben. Den Zusammenhalt habe sehr geschätzt.

Bei meinem ehemaligen Vorgesetzten, Matthias, möchte ich für die Freiräume während meiner Promotionszeit bei Siemens bedanken und bei meinen Vorgesetzten und Kollegen bei Amazon für die große Unterstützung auf der letzten Meile.

Meinen Eltern danke ich dafür, dass sie immer an mich geglaubt und mich unterstützt haben. Auf dem Lebensweg mit ihnen habe ich die Disziplin und das Selbstvertrauen erlangt, um diesen Weg überhaupt einzuschlagen.

Abstract

In today's fast-paced economy, trends are developing quickly and force manufacturers to become flexible and reactive. The production control of flexible manufacturing systems is a relevant component that must go along with the requirements of being flexible in terms of new product variants, new machine skills and reaction to unforeseen events during runtime. This work focuses on developing a reactive job-shop scheduling system for flexible and re-configurable manufacturing systems. Reinforcement Learning approaches are therefore investigated for the concept of multiple agents that control products including transportation and resource allocation. Training strategies are developed to make the solution independent from the number of operations per job and to also generalize to unknown jobs or unseen situations. The concept is set-up to obtain cooperating agents to fulfill a common goal. The solution is evaluated based on requirements that are relevant for the user, however, the major advantage in comparison to conventional methods is the robustness to unforeseen situations during run-time. With the integration concept for flexible manufacturing systems, it is shown that the reactive job-shop scheduling system can be integrated into existing operational technology systems.

Zusammenfassung

Bedürfnisse und Trends verändern sich in der heutigen Zeit schneller als je zuvor. Das hat große Auswirkungen auf die Fertigungswelt und Betreiber sollten daher die Produktion flexibel und reaktiv danach ausrichten. Eine wichtige Komponente spielt dabei die Steuerung des Produktionsablaufs, welche mit neuen Produktvarianten, neuen Maschinenfähigkeiten aber auch unvorhergesehenen Situationen umgehen können sollte. Aus dieser Motivation heraus, liegt der Fokus dieser Arbeit auf der Entwicklung einer reaktiven Job Shop Scheduling Lösung für flexible und re-konfigurierbare Fertigungssysteme. Es wird der Einsatz von Reinforcement Learning untersucht, um mehrere Produkte mit Agenten inklusive Transport und Maschinenzuweisung zu steuern. Dafür werden Trainingsstrategien entwickelt, welche die Lösung unabhängig von der Anzahl an Operationen pro Auftrag macht und die Lösung gleichzeitig so verallgemeinert, dass sie mit unbekannten Aufträgen oder Situationen umgehen kann. Ziel ist es kooperierende Agenten zu trainieren, welche das gemeinsame Ziel einer geringen Fertigungsdauer erreichen. Das entwickelte Konzept erfüllt die Anforderungen, die zusammen mit den Anlagenbetreibern und Experten erarbeitet wurden. Durch die Integration in eine Beispielanlage, wird demonstriert, dass das Integrationskonzept den Einsatz der reaktiven Job Shop Scheduling Lösung in bestehende Systeme ermöglicht.

Contents

Abbreviations

AC	Alternating Current
AGV	Autonomous Guided Vehicles
AI	Artificial Intelligence
AMCC	Avoid Maximum Current Makespan
CAD	Computer-Aided Design
CG	Coordination Graph
CQ-Learning	Coordinating Q-Learning
DDPG	Deep Deterministic Policy Gradients
DHCP	Dynamic Host Configuration Protocol
DP	Dynamic Programming
DPR	Dispatching Priority Rule
DQN	Deep-Q-Network
ERP	Enterprise Resource Planning
ESM	Electronic Supplementary Material
FIFO	first-in-first-out
FJSSP	Flexible Job Shop Scheduling Problem
FMS	Flexible Manufacturing System
HMI	Human Machine Interface
HoQ	House of Quality
ID	Identifier
IDMG	Interaction-Driven Markov Games
IL	Independent Learners
IP	Internet Protocol
JAL	Joint Model Learning
JSSP	Job Shop Scheduling Problem
LPT	Longest Processing Time

MADDPG	Multi Agent Deep Deterministic Policy Gradients
MADQN	Multi Agent Deep-Q-Network
MARL	Multi Agent Reinforcement Learning
MC	Monte Carlo
MDP	Markov Decision Process
MES	Manufacturing Execution System
ML	Machine Learning
MLP	Multilayer Perceptron
MMDP	Multi-Agent Markov Decision Process
MpFJSP	Multi-plants-based MrFJSP
MrFJSP	Multi-resources FJSSP
NN	Neural Network
NP	non-deterministic polynomial-time
OPC UA	Open Platform Communication Unified Architecture
OT	Operational Technology
P-FJSSP	Part FJSSP
PLC	Programmable Logic Controller
PLM	Product Lifecycle Management
POMDP	Partially Observable Markov Decision Process
PPO	Proximal Policy Optimization
QFD	Quality Function Deployment
ReLu	Rectified Linear Unit
RFID	Radio-Frequency Identification
RL	Reinforcement Learning
RQ1	Research Question one
RQ2	Research Question two
RQ3	Research Question three
SAC	Soft Actor-Critic
SCQ	Sparse Cooperative Q-Learning
SGD	Stochastic Gradient Descent
SIE	Importance Satisfaction Evaluation
TD	Temporal Difference
TRPO	Trust Region Policy Optimization
VDMA	Mechanical Engineering Industry Association
WIP	work-in-progress

List of Figures

List of Tables

Introduction

1

Scheduling is highly present in our lives. We are confronted with it in various situations, e.g., when getting the time-table of our children's classes, we receive a well-thought out schedule in which teachers with the skills to teach certain subjects are allocated to specific time slots and classes. When using our smartphones for a phone call, our voice is sent via Internet Protocol (IP) by packages that have to be properly scheduled based on the traffic on the line, so that every package arrives on time. In warehouses, the scheduling of customer orders received and the dispatching to the right pod are indispensable. In manufacturing systems, the inter-dependencies between the involved parties become even more complicated. Flexible Manufacturing Systems (FMSs) are often complex constructs consisting of manufacturing modules that involve machines or handling systems, transportation systems, and control software (Chryssolouris, 2013). FMSs can produce a large variety of products using different machines with partly overlapping skill-sets. Each operation to manufacture a product is dispatched by the production control system to one of many machines with an emphasis on different objectives, such as processing time, energy consumption, and material costs to achieve maximum efficiency.

Manufacturers and retailers are strongly influenced by the market and must align their organizations in such a way that they are flexible enough to react to the consumer's needs at any time. With every advancement, for example, one-day ground delivery, customers' expectations of a seamless and comfortable purchase and shipping experience grow (Ian MacKenzie and Noble, 2013), which in turn is a market pull for the companies. Figure 1.1 shows the interaction between the external environment[1] of a company, such as market demands and new developments, and their organization. In today's fast-paced economies, trends change quickly, leading

---

[1] The environment classification in figure 1.1 is derived from Kirwan (2000), while the influences are derived from Huber (2016)

S. Bär, *Generic Multi-Agent Reinforcement Learning Approach for Flexible Job-Shop Scheduling*, https://doi.org/10.1007/978-3-658-39179-9_1

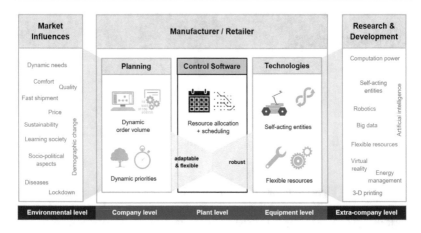

Figure 1.1 The external environment (market demands, research and development) influences manufacturers and retailers. The affected internal environment (company's organization) requires the control software to be adaptable, flexible, and robust

to high dynamics in customer needs and order volumes. Major socio-ecological impacts, such as climate change or pandemics can lead to a modified way of living with changed priorities. Modern FMSs are strongly affected by this and can only cope with the fast-paced economy by utilizing proper control software that is both flexible in terms of varying order volumes, and adaptable to new product variants. Furthermore, large-scale warehouses and FMSs are constantly evolving in terms of their applied technologies, which forces the production control to be robust to the accompanying uncertainties that can arise. Self-planning Autonomous Guided Vehicles (AGVs) for example, greatly increase the flexibility of transportation, while simultaneously increasing the level of uncertainty resulting from deviations in arrival times. The additional requirements that can be derived in this context and how they influence both aforementioned industries are described in detail in chapter 2.

Generally, the dispatching of jobs (orders) to resources (machines) is defined as the nondeterministic polynomial-time (NP)-hard Job Shop Scheduling Problem (JSSP) (Garey et al., 1976), which means that no algorithm can find the optimal solution for polynomial time. Conventional production control systems perform the job prioritization and machine dispatching offline before production starts and conduct a re-scheduling in case of deviations. Flexible control software for FMSs can be split into two major components. Before production starts, there is a prioritization of production orders that must be processed within a finite period, i.e., on the day in question. Once the orders are released to the plant, the reactive scheduling system fulfills the dispatching of orders to machines and transportation decisions online during the production phase (Schmitt et al., 2012). The focus of this work lies on the reactive component of the control software, namely the reactive scheduling system. While operations research has shown great progress in scheduling solutions to solve the JSSP during the past decades and it is still an actively researched field, the potential for improvement grows with advances in algorithms and computational resources being made available to solve the complex optimization problem (Zhang et al., 2019). Well-engineered solutions are available, although these are often specifically tailored to a certain problem. The branch-and-bound heuristic (Brucker et al., 1994) for example tries to find the exact solution, which requires huge computational efforts over and above the extensive engineering efforts. Therefore, it is legitimate to approach the solution with near-optimal methods, such as meta-heuristics, e.g., Tabu-search (Nowicki and Smutnicki, 1996). Dispatching rules (Panwalkar and Iskander, 1977) are simple rule-based approaches that calculate prioritization based on the optimization objectives and prioritize the most urgent jobs. The balance between a solution with high engineering effort and one that is too inaccurate must be found. Modern approaches involve Machine Learning (ML), and more specifically Reinforcement Learning (RL) techniques to develop a low-effort scheduling approach with the advantage of fast reaction during the operation phase resulting from less computational effort (Gabel, 2009) (Waschneck et al., 2018) (Kuhnle et al., 2020) with the drawback of not being easily adaptable to different machines or products.

The contribution of this work is to develop a self-learning approach that involves less engineering effort for adaptation and scaling. The scheduling solution should meet the requirements of being reactive during production to a relevant unforeseen situation, must handle machines with multiple and overlapping skill-sets and should schedule a large number of product variants. The solution is tailored for job-shops or conveyor systems with multiple distributed machines in which manufacturing processes (drilling, milling, grinding), assembly processes (assembling components to a base plate) and other value-adding or quality-related processes are required. It is especially suitable for systems with high dynamics and a possibility for re-scheduling, or in other words, systems where online decision-making is desirable. A rough offline plan can be created in advance using an offline planning solution to prioritize the orders and determine the release times to the system, and our system should then find the optimal sequence of released jobs while considering plant constraints and minimizing the transportation time. Consequently, the customer would be the manufacturing system's operator, which can either be the operations manager or site leader, depending on the size and organization of a company. The operator's requirement is to run a production with high outcomes while being able to cope with their vendors' or end-customers' demands.

From the identified need, the overall research goal of this work is derived: **How can a concept of reactive scheduling meet the demands of FMSs?** In section 1.1, we explain *how* this overall goal is distributed in research questions, followed by the applied methodology in section 1.2 and the structure of the work in section 1.3.

1.1   Research Goals

The overall goal of the current work is to design a concept of reactive job-shop scheduling to meet the requirements of FMSs. The aim is to investigate how a self-learning approach can be used to find good schedules. The solution should focus on machine dispatching and transport decisions of already released production orders, which implies that the a-priori order prioritization is not part of the concept. It will be evaluated whether the self-learning and reactive production scheduling approach ensures high levels of productivity in unforeseen situations while taking advantage of the flexible resource capacities offered by the FMS.

Research Question one (RQ1)—How can the extended job-shop scheduling problem be formalized and the concept be designed to fulfill the requirements of reactive production scheduling? A suitable survey process is used to identify the requirements of current FMSs for job-shop scheduling problems. The answer to this question involves the formalization of the extended problem and the concept of self-learning entities that are trained to facilitate the desired reactive production scheduling system.

Research Question two (RQ2)—What does a suitable training set-up look like, including the selection of algorithms, development of the training environment, and the training concept itself, to achieve a common optimization objective within reactive production scheduling? The answer to this central research question includes the selection and the development of an enhanced concept of an applied self-learning algorithm, in addition to the training strategy to enable efficient learning and cooperative behavior of entities. The training concept must consider how to handle local and common optimization objectives and should be scalable to high production variances. The comparison of the different training approaches with respect to valid and near-optimal schedules and the evaluation based on the requirements concludes this research question.

Research Question three (RQ3)—How can the deduced solution be integrated into an existing real-world manufacturing system? This research question focuses on the integration concept to real-world manufacturing systems and evaluates the hurdles involved in deploying this solution in a brownfield system, including the integration into existing Operational Technology (OT) systems. Through this, the limitations of this solution are discussed and it is shown where further research could connect.

1.2 Methodology

The Quality Function Deployment (QFD) is used as methodology to approach the presented research questions. QFD is usually applied in product development, quality management, and technology planning and is an approved method to develop quality functionalities required by the customers according to the standard DIN EN ISO 9001. Certain steps of the QFD are selected and adjusted to guide this research instead of using it for product development. The full concept can be found in Rabl (2009), while the adjusted and used parts are described in the following. We only consider the first step of the QFD that has the goal to develop a solution that features the customer's requirements. Generally speaking, this is done by bringing together the two perspectives: market view and technical view. The market view helps to focus on the market with its potential customers and competitors, while the technical view then helps to design the solution with its functionalities in a way that considers the requirements that were defined by the market view.

The House of Quality (HoQ) (Rabl, 2009) is the tool that is used in the first step of the QFD, which is the method used to develop the reactive scheduling system to answer the research questions. The HoQ helps to perform the necessary steps and to maintain the overview of the structure of the thesis. It is depicted in figure 1.2 with the ten steps that first help to determine WHAT should be solved, WHY it is needed, and lastly HOW the identified problems can be solved. The market view contains steps 1 to 4, in which the customer requirements are listed and weighted, and alternative publications are reviewed and evaluated based on the extent to which they already fulfill the customer requirements. In step 4 it is therefore determined where further research is needed. This is the pre-requirement to start with the technical view that consists of steps 5 to 10. From the defined research gap, in step 6 the technical functionalities can be determined that are essential for coping with the defined requirements of step 2. Dependencies and conflicts between these functionalities are specified in the roof of the HoQ in step 7. Step 8 helps to validate whether the defined functionalities can handle all customer requirements and, lastly, the results of the developed solution are evaluated based on these in step 9. Step 10 was added to derive the potential and strategy for improvement of the solution from the evaluation. The detailed description of every step and the relevant results, can be found in the respective chapter where it is applied. The structure of the thesis including the HoQ is presented in figure 1.2.

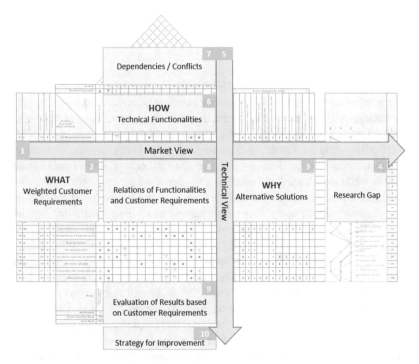

Figure 1.2 HoQ with ten steps for developing a solution that is relevant to the customer and not yet provided by other research. The market view helps to understand the customer requirements in steps 1–4, while the technical view defines and evaluates functionalities based on the requirements in steps 5–10 (based on Rabl (2009))

1.3   Structure of the Thesis

This thesis is structured in nine chapters and addresses the three research questions that are presented in section 1.1. The HoQ is the guiding principle that runs through the work with ten adjusted steps that are performed in different chapters as indicated in figure 1.3 on the left-hand side. The market view was briefly addressed in the introduction to demonstrate the influences on FMSs and is further elaborated in chapter 2 to motivate why flexible production scheduling is necessary and which customer requirements need to be considered as the first step of the HoQ. In chapter 2, we formally introduce the JSSP, followed by step 2 to obtain the ranked customer requirements for production scheduling in an FMS. The first part of the state of the art is presented while using step 3 of the HoQ as a structured approach to analyze alternative publications that aim to solve production scheduling challenges with self-learning algorithms. In step 4, it is subsequently determined where further research is needed to cope with the defined requirements. Chapter 3 gives an introduction to RL, which is the applied method for developing a self-learning reactive scheduling system with less engineering effort. It involves Multi Agent Reinforcement Learning (MARL) approaches that are used to solve similar complex decision-making problems. These approaches are reviewed and evaluated to determine which of them can be adjusted and applied to the problem at hand. The resulting concept and formalization of a self-learning reactive scheduling system with all technical functionalities and dependencies (steps 6–8) is subsequently presented in chapter 4 to answer RQ1.

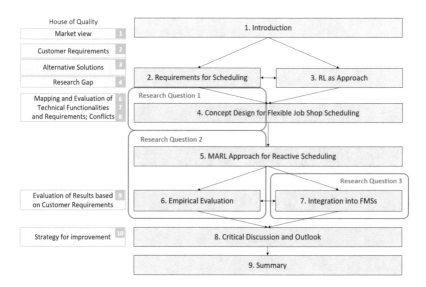

Figure 1.3 Structure of the thesis

With the concept resulting from RQ1, in chapter 5 different solution designs are presented for achieving cooperative behavior of self-learning entities to find good schedules. To compare the proposed solution designs, it is necessary to rate them based on valid and optimal schedules. The most promising solution design is evaluated using step 9 of the HoQ based on the criteria defined in the market view. With this evaluation in chapter 6, the second and central research question is concluded. Following chapter 5 and chapter 6, in chapter 7 it is demonstrated how the solution can be integrated into a real FMS. The applicability of the solution and the integration concept into existing Manufacturing Execution Systems (MESs) is evaluated by deploying the scheduling system to an example FMS to answer RQ3. Chapter 8 discusses the limitations the solution faces with respect to the derived requirements by performing the last step of the HoQ. It is further pointed out where future research could connect. This work is briefly summarized in chapter 9.

Requirements for Production Scheduling in Flexible Manufacturing

In this chapter, we elaborate the requirements for flexible scheduling systems in more detail, analyze state of the art scheduling solutions, and differentiate our solution from these. To obtain a general understanding of JSSP, a formal introduction is provided in section 2.1. We then analyze the requirements needed for advanced scheduling solutions to fulfill the needs of modern and flexible production by applying an adjusted version of the HoQ. The first step in the HoQ is the collection of requirements and the second step is to weight them according to importance from the customer's point of view, which is described in section 2.2. In the third step, alternative solutions are considered and rated according to how well they fulfill the identified requirements in section 2.3. Further, the HoQ helps to clearly identify the research gap (step four) in the alternative solutions in section 2.4, which enables us to differentiate our work and is also a preparation for defining relevant functionalities. Finally, the extended JSSP for flexible manufacturing is specified in section 2.5.

2.1 Foundations of Flexible Job-Shop Scheduling Problems

Producing a product within a manufacturing system or job-shop involves a sequence of processes. Considering an FMS with simplified assumptions, the scheduling of these processes can be formalized as classical JSSP. In job-shop scheduling n jobs or tasks must be assigned to m resources that are either machines or manufacturing modules. One of the most important characteristics of the JSSP is that the order of sequences is pre-defined. A relaxation of the JSSP is the Flexible Job Shop Scheduling Problem (FJSSP) that was introduced by Brandimarte (1993) with the addition that machines can have overlapping skill-sets. In FJJSP, a job j consists of i_j operations $o_{1,j}, ..., o_{i_j,j}$ from which each can be processed either on a sub-set of resources $\sigma(o_{k,j})$ (partial flexibility) or on all resources (full flexibility) within the expected processing time $\delta(o_{k,j})$. The completion time c_j determines the time a job needs to be completed. The FJSSP is defined by a tuple $J = [\mathcal{J}, \mathcal{M}, \mathcal{O}, v, \sigma, \delta, \tau]$ with $\mathcal{J} = j_1, ... j_n$ as the set of jobs, $\mathcal{M} = m_1, ..., m_m$ as the set of machines, and $\mathcal{O}$ as the set of operations per job. The function $v\colon \mathcal{O} \to \mathcal{J}$ specifies the operations of the relevant job. The function $\sigma\colon \mathcal{O} \to \mathcal{M}$ defines which operation can be processed on which machine and through the function $\sigma'\colon \mathcal{O} \to \mathcal{M}$ it is determined which operation is processed on which machine that needs to be found. Function $\delta\colon \mathcal{O} \to t$ gives the time steps needed to execute each operation. And finally, τ defines the order of sequences in which the operations must be executed. Figure 2.1 visually depicts the introduced tuple, including the searched function σ' that defines the machine dispatching, exemplary for the first operation.

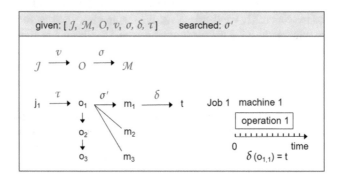

Figure 2.1 The tuple J is given to describe the JSSP. Function σ' is searched to define the schedule. In this example, it is determined that operation o_1 of job j_1 should be processed on machine m_1 at time 0 with processing time t

We allow a product to recirculate within the manufacturing system and thereby to use machines multiple times for different operations. When solving the FJJSP, it is desirable to find $\sigma'(o_{k,j})$ for all $k \in [1, ..., m]$ and $j \in [1, ..., n]$, so that an objective function is optimized. Objective functions are performance measures that evaluate the solution based on the desired goal. An objective function can be time-based, when it is requested to finish jobs fast, it can aim to reduce work-in-progress (WIP) to ensure a high throughput or simply to finish jobs before their due dates. Multi-objective approaches consider the combination of different objectives. In JJSP it is commonly aimed to minimize the total makespan C_{max} of the jobs as considered by Csáji and Monostori (2008):

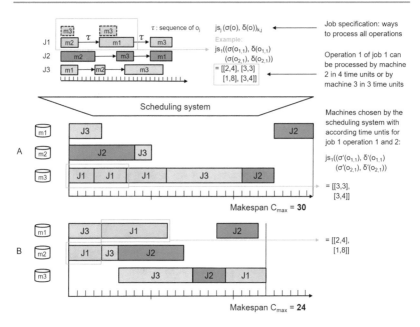

Figure 2.2 Three job specifications are given, where job 1 (J1) has flexible options in the first and second operation. A comparison of schedules A and B demonstrate the impact of optimal scheduling with respect to the total makespan. The transportation times are assumed to be constant but are not displayed

$$C_{max} = \max_{j=1,\dots,n} c_j, \tag{2.1}$$

that is defined by the time in which the last operation $o_{i_j,j}$ of the last job j of a sub-set of jobs has been processed. In figure 2.2 we demonstrate the impact of proper scheduling concerning the total makespan C_{max}. Three jobs or products are considered with a given order of sequences of operations τ that need to be processed. For each job, $\sigma(o_{k,j})$ defines on which machine each operation *can* be processed. Job 1 is the only one that comes with some flexibility with regard to operations 1 and 2, as operation 1 can be processed by machine 2 or 3. Function $\delta(o_{k,j})$ defines the time steps needed for each possible machine $\sigma(o_{k,j})$. The job specification js is the formal description of how all operations can be processed with the relevant properties and is defined by $js_j(\sigma(o), \delta(o))_{k,j}$.

The two possible schedules in A and B are presented using a common approach to visualize resources and scheduled items known as a Gantt chart. The x-axis represents the time and the y-axis the machines. Each operation of each job is depicted in the line of the given machine where it is processed, with the start and end time according to the x-axis. The duration of each operation is thereby clearly visualized. The schedules shown in A and B are defined by $\sigma'(o_{k,j})$ and $\delta'(o_{k,j})$ and emphasize the considerable impact in regards to the makespan of using both alternative machines for job 1. When all three operations of job 1 are processed by machine 3 in schedule A, it is beneficial for job 1 to finish early, although this leads to a waiting time for job 2 until the second operation can be started. Thus, even though job 1 finishes very fast, the total makespan suffers. In schedule B, job 1 was the one that had the longest processing time leading to an overall smaller total makespan. Considering C_{max} as the global optimization objective, it is obvious that schedule B is the better solution in this case. This indicates already that jobs being produced in parallel must be scheduled wisely and entities controlling these jobs for the scheduling should consider each other and cooperate to achieve the global goal of minimizing the makespan. The complexity of FJJSP for FMS is discussed in more detail in subsection 2.2, where the requirements are derived from two concrete examples.

2.2   Requirement Analysis of Flexible Scheduling Solutions

In chapter 1, we touched on the overall influences that drive manufacturing and scheduling systems to become more scalable, flexible, and robust. The goal of this section is to analyze what this means for a proper scheduling system and which requirements can be derived from the challenges and desires for FMSs. As described, we use the HoQ as a methodology to set the right focus in the early phase of the research and describe the results from the first and second steps in this section to delimit the requirements we want to consider in our solution design. Therefore, we provide two concrete examples from the warehouse industry in section 2.2.1 and the manufacturing industry in section 2.2.2 in which requirements are introduced with (IDs) that we summarized in table 2.1 and ranked by the operators and domain experts in section 2.2.3.

2.2.1   Influences on Warehouse Control Systems

After years of research in the field of robots and multi-agent systems, a disruptive material handling system was first deployed in the warehouse industry in 2006. AGVs were used since the 1950s to move heavy objects in warehouses, but

Table 2.1 Requirements for production scheduling with IDs derived from the warehouse and manufacturing examples

ID	Requirement	ID	Requirement
12	Huge amount of product variants	20	Low effort in adaptation
13	New products with less effort	7	Transport decisions
16	Overlapping skill-sets of machines	19	Multiple skills per machine
14	New machine skills with less effort	4	Proactive solution
17	Multi-objective optimization	8	Flexible machine topologies
18	Global optimization goal	21	Easy transferability
6	Reaction to unknown situations	9	Transportation times
5	Reaction to unforeseen situations	3	Near-real-time decision-making
11	Complex material flow	10	Various means of transport
15	Less engineering effort		

with advances in wireless communication for navigation, computational power, and advances in the vehicles themselves, the use of AGVs in warehouses became state of the art (Poudel, 2013). Mobile robots (drives) lift movable storages (inventory pods) from a storage location and bring them to the workers (pickers), who pick the displayed items to process customer orders as shown in figure 2.3. The productivity of pickers has increased by a factor of six when compared to a classical man-to-good and pick-to-cart environment (Chain, 2021). The rates increased up to 400 units per hour as a result of the fact that pickers stay at their stations, while drives continuously present shelf-faces to the pickers. The Kiva warehouse management system, today known as Amazon Robotics (Robotics, 2015), is a new paradigm for pick-pack-ship warehouses introduced by Wurman et al. (2008). With a significant improvement of the overall productivity as a business case with a return on investments of one to three years (Wurman et al., 2008), it became the state-of-the-art in modern warehouses (Poudel, 2013). This advance makes it possible for online retail businesses to react to customer demands within hours and fulfill customer orders within one working day. The objective of the control system is to meet each delivery promise while minimizing the makespan that is defined by the time when the first drive starts with its job until the last drive finishes its job in a defined set of jobs (17, 18). In this warehouse scheduling example, one job usually consists of only one task, namely the transportation of the right pod to a worker that has the eligibility to pick the customer order. The makespan can thus be minimized by selecting pods and drives that avoid traffic (10) wisely and proactively (4).

The following requirements are found in a fast-paced warehouse environment and are fulfilled by the Kiva solution (Poudel, 2013):

- Fast decision-making (3): customer orders that are received are processed within minutes with a fast reaction to the dynamic environment, e.g., varying numbers of workers (14), obstructions on the floor (5), or drives that are out of service.
- Easy to adapt (20): the number of pods, drives, and usage of stations can be ramped up for more capacity according to seasonal business needs (8, 14).
- Flexible resources (8): workers with different eligibilities can sign in on any station and receive proper tasks.

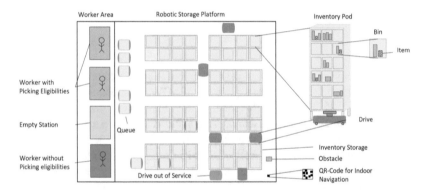

Figure 2.3 Illustration of a robotic storage platform in a pick-pack-ship warehouse. Drives lift pods and bring them to the workers with eligibilities to pick the items

It is worth mentioning that one item type can be stored multiple times in one pod or bin, which is highlighted by the shaded item in figure 2.3 and it can be further stored in multiple inventory pods, making the decision of which pod to choose even more advanced. The storage logic in general follows the chaotic storage system, but pods with frequently used items are usually located near the work stations. From the aforementioned requirements, it can be derived that the control software of the system must be flexible in terms of the available resources and must react online to any changes. Kiva's approach is to distribute the complex control problem to sub-tasks and solve it with a multi-agent system, in which decentralized agents control online tasks. This decision was made for the following reasons:

- The optimization problem is defined by the combination of tens of thousands of customer orders, tens of thousands of pods, and thousands of drives. Currently, the complete and optimal solution for the high-dimensional problem cannot be calculated fast enough with a reasonable amount of computation power. Therefore, the approach is to use different agent types for relevant tasks that approach a good solution, e.g., the selection of the drive that should be used for the current job based on partial observations of the environment, instead of calculating the complete schedule of tens of thousands of customer orders in advance.
- In the fast-paced online retail business with same-day delivery promises, there is a need for real-time and reactive decision-making because customer orders are received continuously and must constantly be re-prioritized to consider the critical pull times of trucks shipping the parcels to distribution centers. It would be impractical to consider a frozen period of customer orders that are scheduled offline as is common in the automotive industry with production times of several months. Therefore, decentralized agents perform their decision-making online and in a proactive and reactive way.
- The dynamic environment is dependent on the state of other drives, traffic on the routes, obstacles such as dropped items on the floor, managed areas for maintenance workers, and the fluctuating headcount of workers at the stations depending on the buffer situation. Therefore a re-calculation is constantly needed based on the dynamics of the environment.

With the design decision of a multi-agent system, the complexity of the problem is distributed to multiple decision-makers and includes the aspect of cooperation. The problem is split into task that are distributed to manageable components with partial observations, but clear interfaces and responsibilities (Lesser, 1999). Cooperation can be indirectly achieved by sharing information or experiences with other agents or by direct communication via a protocol. Overall, there are three agent types with different scopes of responsibilities. The first agent type is the overall job management software that receives, prioritizes, and dispatches customer orders to pods, drives, and stations and is therefore responsible for scheduling and resource allocation (Wurman et al., 2008). Each pick task only consists of one process that needs to be scheduled, namely picking the customer order, but other tasks that can be assigned to the same pod must also be taken into account in the job manager's decision-making. In addition, tasks can have different priorities that must be considered, depending on the critical pull time needed for shipping.

Each station and drive is represented by an instance of the station or drive agent respectively, which both act and cooperate via XML messages in a controlled and known environment. The inventory station agent is instructed by the job manager to ask drive X to bring pod Y and show face Z. It sub-sequentially provides the relevant rack lights to navigate the picker to the item, reports the accomplishment of tasks to other agents, and releases the pod if there are no further items that can be picked from this pod. One drive is dispatched to one pod with the individual goal to fulfill a customer order. The drive agents plan their path to fulfill their assigned task, navigate using fiducial markers on the floor, and circumvent obstacles using sensors (Wurman et al., 2008) (Poudel, 2013). Each drive is independent of other drives but should consider other drives to fulfill the common goal of low traffic to achieve a minimum makespan. In a scenario in which a pod is assigned to multiple stations, the drive needs to find a path with low costs in terms of travel and waiting times, which is similar to the Travel Sales Man Problem. This is an NP-hard problem (Garey et al., 1976), which means that the optimal solution cannot be found in polynomial time, but can be approached by heuristics such as the nearest neighbor (Hart et al., 1968) or Dijkstra's algorithm (Vahdati et al., 2009), which is used by the Kiva system (Poudel, 2013). To summarize, the warehouse management system consists of two major optimization problems, namely the resource allocation solved by the job manager and the path planning fulfilled by the drive agents online on the floor. Even though the variance of tasks assigned by the job manager in the warehouse management is low if we compare it to production processes in the industrial manufacturing world, the complexity of the resource allocation problem is high and very similar to the job-shop scheduling problem in production systems where jobs have multiple tasks that must be processed on different machines. In general, there are many parallels between the warehouse management problem and the job-shop scheduling problem in the manufacturing world.

2.2.2   Influences on Manufacturing Control Systems

The development of manufacturing systems is also driven by the continuous deployment of advanced automation techniques and new technologies. The final transition from mass production to mass customization requires manufacturing systems to be flexible, adaptable, and cost efficient (Zacharia and Xidias, 2020). According to the definition of Umar et al. (2015), a manufacturing system becomes an FMS when it has the ability to produce a variety (12) of products in parallel by automated machines that can handle different tasks (19). One task can also be processed by different machines as a result of the overlapping skills of the machines (16). It must also consist of a flexible transportation system and is further distinguished by an adaptable production control system that can deal with changed priorities and dynamic job releases (Umar et al., 2015). Conventional manufacturing systems where mass

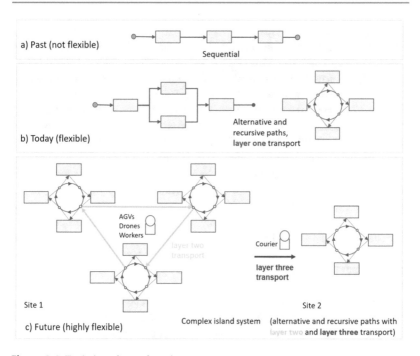

Figure 2.4 Evolution of manufacturing systems

production is in focus have one common goal: productivity. Most of these systems have dedicated manufacturing lines with a sequential structure where every product must pass every machine in a pre-defined order for value-adding activities, as shown in figure 2.4a. However, when striving for fast and efficient production, a modified plant structure is advantageous (Huber, 2016). Alternative paths enable more variance in material flow and a circular plant topology (11) enables products to return to machines at a later time, as shown in figure 2.4b. Concepts such as that proposed by Melfi (2015) consider production islands in the job-shop connected by AGVs or drones, as indicated in figure 2.4c, with multiple means of transportation (10) on the one hand and additional uncertainties (5) due to unpredictable transportation times on the other hand. We categorize transportation into three main layers:

- Layer 1: transportation within a production line (e.g., conveyor belts),
- Layer 2: transportation between production lines/machines (e.g., AGVs, humans), and
- Layer 3: transportation between sites (e.g., trucks, ships, planes).

From layer to layer, the impact of delays in transport times increases. When the arrival time of anAGV deviates because of a necessary re-routing the affected WIPs can be processed late. The delay of a truck due to traffic on the road can have even bigger effects on the dependent production system waiting for the material. Scheduling systems are directly dependent on transportation, which means that deviations in terms of the arriving product or material can disrupt the whole plan. When optimizing the utilization of machines, the transportation between different machines plays a crucial role (7) (Huber, 2016). Information about the transportation costs between all locations, which for example is defined by the average time units need for a given transportation section, must be known and considered in the decision-making process (9). If there is more than one means of transportation available for a certain section, this should also be taken into account (10). The focus is thereby kept on optimizing the intra-plant logistic process including the layer 1 and 2 transport, but we exclude layer 3 transport that goes more in the direction of the general supply chain. We assume that the proper planning of layer 3 transport, including backup plans for delayed arrivals, is considered by a superior planning system that our solution can rely on.

To strive for mass-customization on the one hand and to cope with fluctuating product demands on the other hand, it is efficient to utilize modular machine topologies (8). With this, manufacturing modules are plugged around the transportation system and can easily be exchanged when producing a new product variant (13) (Koren et al., 1999), (Chan and Zhang, 2001). Flexible module topologies and the accompanying dynamics of available manufacturing modules must be further handled by the scheduling system (Melfi, 2015), (Shaik et al., 2015). This includes the dispatching of orders to any of the machines within the job-shop to which transport is currently available, independently of where the WIP product is located. Besides applying a modular machine topology, it is advantageous to utilize machines that can be reconfigured and can have more than one skill (19) in order to meet the high frequency of product innovations due to regulatory requirements or market demands (Hurink et al., 1994), (Umar et al., 2015). A skill of a machine is defined by its ability to perform an action in a defined manner, e.g., drilling a hole with a certain speed and depth or placing a certain element in a certain position. A machine with multiple skills would, for example, be able to place one element out of five different types of material in five different positions on a workpiece, resulting in 25 different skills if the machine is provided with the five types of material. If one material is out of stock, the five skills to place the missing material in the five positions would be unavailable until the material gets refilled. Autonomous mobile robots, for example can have transport skills, grasping skills, and placement skills. In modern factories, it is common that machines have overlapping skill-sets (16) to introduce even more flexibility in case a certain skill is required very often by the products (Umar et al., 2015). To quickly react to upcoming product advancements, machines can easily be reconfigured on demand by a new tool or material, thereby rapidly introducing new skills (14).

Adequate production control is one key element in the enhancement of efficient mass customization and must go along with the challenges that arise. One example is the overall fuzziness that can result from customized product orders because the customer has the option to create products that have never been produced before and are therefore unknown for the production control system (6). The job specification that describes how the product can be manufactured is generated with the receipt of the customer order and the expected production time is calculated based on rough estimations because the actual times are unknown to the system. Additional deviations can arise due to machine down-times, which requires a proper control system to immediately reschedule the WIP products to alternative machines (5) to avoid traffic and aging WIPs (Souier et al., 2019). From these exemplary scenarios it can be derived that in FMSs it is hardly possible to create an exact offline schedule, as the actual schedule is likely to deviate from the plan. Even when manufacturing the exact same product, the machine dispatching can deviate from the last time due to different context information including that of other products and the demand of machines. This leads to the requirement of having a flexible production control that is capable of online scheduling or a fast re-scheduling of all products (3). This motivated Schmitt et al. (2012) to propose a hybrid approach of planning- and value-oriented production control. Planning-oriented approaches are successful in enterprises where an exact model of the production system is given and a defined set of customer orders need to be prioritized (e.g., based on due dates), allocated to resources, and released to the FMS in the defined order. For every deviation or change in priorities, a complete re-planning must be executed that is both time-consuming and cost-intensive. To achieve more flexibility, central planning is re-assigned to the value-added processes in the value-oriented approach, meaning that decision-making is executed decentrally and situation-based. A hybrid approach of the a-priori planning-oriented and the value-oriented production control is the favored concept of Schmitt et al. (2012), where the prioritization of customer orders is performed before production starts and resource allocation and path planning are executed online during production, which is in fact similar to the approach used in the Kiva warehouse management system. One realistic concept of the intra-organizational supply chain in manufacturing industries is shown in figure 2.5, where the planning time horizon and its purposes are inspired by (Zhao, 2020), and the schematic follows (Foit et al., 2020) and is further based on chapter 5 of (Thomas et al., 2006).

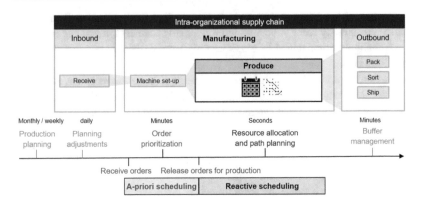

Figure 2.5 Production planning and scheduling of customer orders within the planning time-line of the intra-organizational supply chain of manufacturing systems

In general, the supply chain within a factory consists of inbound where raw material and goods are received, the manufacturing process itself, and outbound where the finalized products are packed, sorted, and shipped. A monthly planning is usually conducted for predicting production activities and commissioning of raw and ancillary materials and tools based on the predicted order volumes, followed by a weekly planning to adjust the plan when necessary. After receiving the actual goods, daily adjustments are made regarding the workforce and machine set-up with respect to the continuously received customer orders that are prioritized according to specific criteria such as due dates, first-in-first-out (FIFO), or available resources by the a-priori scheduling system. This planning is based on defined processing times of the production system and assumes ideal conditions. The prioritized orders are sub-sequentially released to the production system. From this point in time, the reactive scheduling system is responsible to allocate resources for each job and perform the path planning on the flexible transportation system based on the actual situation in the production system (Schmitt et al., 2012). The finalized product is sent to outbound, where the buffer must be maintained that in turn affects the production capacity, e.g., a full buffer means that the throughput of production must decrease.

After comparing this intra-organizational manufacturing concept with the example of the control system in large-scale warehouses in section 2.2.1, the parallels identified are opposed in table 2.2. In the past, it was analyzed how advancements that were applied in warehouses, such as AGVs, were later applied to the manufacturing industry, so it makes sense to also analyze the advances of the control system in that business. For the criteria resources, material flow, and uncertainties, both domains deal with similar circumstances. However, in the warehouse industry, one job mainly consists of only one task that needs to be prioritized and dispatched to a resource, while in the manufacturing industry one job often consists of multiple tasks that need to be processed by different resources, respectively machines, which adds variance and increases complexity. The control system in warehouses is hybrid consisting of an a-priori job prioritization, in addition to resource allocation and online path planning components executed by decentralized drive agents. In the manufacturing industry, it is also observed that the trend goes in the direction of having an a-priori job prioritization that can consider resource allocation but which mainly relies on the online resource allocation and path planning system during production (Schmitt et al., 2012).

Table 2.2 Comparison of warehouse and manufacturing control

	Warehouse	**Manufacturer**
Resources	Worker with eligibilities	Machine with skills
Material flow	Transportation of goods	Transportation of WIP products
Uncertainties	Arrival times, headcount change	Arrival times, machine failures
Job complexity	One task	Multiple tasks
Control system	Hybrid: a-priori job prioritization and resource allocation, reactive path planning	Hybrid: a-priori job prioritization, reactive resource allocation and path planning

2.2.3   Derived and Ranked Requirements

From the two examples given, we summarize and complete the requirements that must be considered by the reactive scheduling system to further rank them. The requirements that were found in the warehouse example are mapped to requirements addressing the scheduling solution for the manufacturing industry and named accordingly.

Complex material flow (11) and modular machine topologies (8) are characteristics of FMS and require the scheduling solution to consider the transportation itself (7), including various means of transport (10) and the transportation times (7) in the decision-making. A flexible scheduling system must quickly handle multiple skills of machines with dynamic availability (19), overlapping skills of machines (16), and new skills of machines (14). Vendors often offer a wide range of product variants and must react to socio-ecological influences by customizing their products regularly. Scheduling systems must therefore handle a large number of product variants (12) including new and completely different products (13) that can vary in terms of the number of operations that must be performed to manufacture the product. Benchmark sets, such as those of Hurink et al. (1994) already consider jobs with various numbers of tasks and overlapping machine skill-sets to evaluate Tabu search on flexible job-shop scheduling. Advanced technologies and high digitization are accompanied by unforeseen situations (5) such as delayed arrival times of AGVs due to unpredictable obstacles on the route, unavailable machines, or conveyance problems. Reactive scheduling systems must be able to deal with these uncertainties and quickly react within seconds (3) (Souier et al., 2019) in addition to being proactive to avoid critical situations (4). The same applies to situations that have never occurred before, e.g., because of arbitrary combinations of product variants that should be manufactured. The solution should be able to deal with any hitherto unseen or exotic situation without waiting time or even the need of human interaction with the system (6). Software solutions scale successfully when they are generic enough to be easily reused and adapted to similar problems. This also applies to production scheduling systems, as manufacturing operators often need to react to their customers' requirements and must adapt their machines accordingly (Csáji and Monostori, 2008). For production scheduling, this means that an easy adaption with less engineering effort (20) is one further requirement to cope with the fast-paced environment in the flexible manufacturing world. There are many scheduling solutions that solve sophisticated problems with highly engineered systems. In this work, we aim to develop a solution that can be used for a new problem with less engineering effort (15).

The collected requirements should be ranked according to the second step of the HoQ, within a range, e.g., from 1 to 19 with high values for high importance. The recommended approach is to obtain this information by conducting interviews with the customers directly but due to the lack of access to real customers, interviews were conducted with 12 experts in the field of flexible manufacturing instead. These experts have different roles ranging from project lead, technical project lead, scheduling expert, researcher, plant operator to head of a research lab for flexible manufacturing. For our purposes, it is legitimate to use this shortcut approach as this step is essential but not the main aspect of this work. Requirements (1) and (2) are neglected in the ranking as they involve given information, namely that considering a job-shop manufacturing system (1) and machine dispatching (2). Table 2.3 summarizes the requirements for production scheduling systems with IDs and the average rankings. Requirements that were ranked in the top eight included: considering a large number as well as new product variants, overlapping and new machine skill-sets, global and multiple local optimization goals, and the reaction to unknown and unforeseen situations. The requirements regarding the consideration of complex material flows and less engineering effort for the solution made it to the top ten.

Table 2.3 Requirements for production scheduling with IDs and relevance for the customer

ID	↑↓	Requirement	ID	↑↓	Requirement
12	15.13	Huge amount of product variants	20	9.88	Low effort in adaptation
13	14.50	New products with less effort	7	9.25	Transport decisions
16	12.63	Overlapping skill-sets of machines	19	9.13	Multiple skills per machine
14	12.00	New machine skills with less effort	4	9.00	Proactive solution
17	11.38	Multi-objective optimization	8	8.63	Flexible machine topologies
18	11.13	Global optimization goal	21	8.38	Easy transferability
6	11.00	Reaction to unknown situations	9	7.50	Transportation times
5	10.88	Reaction to unforeseen situations	3	4.75	Near-real-time decision-making
11	10.38	Complex material flow	10	4.71	Various means of transport
15	10.13	Less engineering effort			

We indicated why it is necessary to adjust the control system within FMSs to deal with the dynamic requirements that arise from the market side or from technology pulls. It was derived that the control system must be adaptable, flexible, and robust to cope with these dynamics. As the classical JSSP is already an NP-hard (Garey et al., 1976) problem, the collected requirements make the problem even more challenging. When evaluating the state of the art, it can be seen that many algorithms aim to find a good rather than an optimal solution. Approximating the optimal solution is thereby a legitimate approach. In the following section, different approaches that are used to solve complex JSSP and FJJSP are described.

2.3   State of the Art: Approaches to Solve Job-Shop Scheduling Problems

Job-Shop Scheduling is an active research field in Operations Research. In general, scheduling can be distinguished according to predictive and reactive solution approaches. Predictive scheduling algorithms search for the solution in advance before production starts. There are approaches that can achieve the optimal solution up to a certain problem size and only if the problem is deterministic, because these algorithms are often computationally intensive and scale exponentially with the complexity of the problem. In contrast, reactive scheduling solutions can react to any intermediate and unforeseen circumstances and feature near-real-time decision-making during production. These can be further split into solutions with rule-based decision-making and more sophisticated solutions featuring proactive situation-based decision-making. In the following section, conventional scheduling solutions and Artificial Intelligence (AI) based RL scheduling solutions are reviewed and their limitations are outlined. We evaluate these based on the requirements defined in section 2.2 to determine the research gap.

2.3.1   Conventional Scheduling Solutions

A classical approach to solve the JSSP is to formalize it as a disjunctive graph (Roy and Sussmann, 1964) to then solve it with exact procedures using search heuristics, such as the branch-and-bound (Brucker et al., 1994). Furthermore, there are near-optimal approaches using meta-heuristics, such as the tabu-search methods (Nowicki and Smutnicki, 1996), beam search (Ow and Morton, 1988), simulated annealing (Van Laarhoven et al., 1992), and the greedy randomized adaptive search procedures (GRASP) (Binato et al., 2002). The idea behind most search algorithms is to iteratively improve the present solution by making small adjustments. Using the cost function, the approach evaluates whether the resulting solution improved as a result of the previous adjustment, i.e., the local search algorithm of Vaessens et al. (1996). While most branch-and-bound approaches are mainly used to solve classical JSSPs with one optimization objective, the enhanced evolutionary algorithm of Esquivel et al. (2002) and the job-oriented heuristic of Hastings and Yeh (1990) can be used for multi-objective optimization of JSSPs. While heuristics are often used to solve classical JSSPs, those needing high computational efforts face limitations when applying them to real-world scenarios, i.e., when they are required to react to any unforeseen (5) or even unknown (6) events with very short reaction times (3). This is why the application to more practical JSSPs was realized by He et al. (1993) with the multiple-pass heuristic that uses a dispatching rule to find a valid schedule that is improved in iterative steps. In Schutten (1998), the shifting-bottleneck heuristic (Adams et al., 1988) was applied to more practical JSSPs involving realistic features such as simultaneous resource demands, transportation (9), and set-up times. The presented heuristic methods require high engineering efforts (15) and adaption for upcoming changes (20) together with the high computational effort and waiting time (3) in unexpected situations, which leads to the conclusion that they are rather impractical for use in the desired flexible and reactive scheduling concept.

However, the Dispatching Priority Rules (DPRs) introduced by Panwalkar and Iskander (1977) are used during production as a reactive scheduling method (5) with the advantage that they are easy to apply (15). DPRs can be used when parts of the problem can be solved in a decentralized way by using only a partial view of the system. One dispatching algorithm is commonly applied to each machine with the goal to assign the next job to the controlled machine. Thereby, local DPRs use information about the availability of the controlled machine, as well as information about waiting jobs in the order stack, and global DPRs also use information about other machines for their decision-making. There are various kinds of DPRs that generally can be classified into static and dynamic ones. Static DPRs basically compute a priority value for each of the jobs in the order stack based on its properties such as the waiting time, due-time, or any other customer-specific optimization requirements (18). Based on these computed values, the rule selects the job that is judged to be the most urgent with respect to the considered optimization objective (Gabel, 2009). There are further dynamic DPRs that change their selection strategy over time, based on their experiences. Wisner (1993) and Pinedo and Hadavi (1992) provide a thorough review of commonly used DPRs in industrial environments. Well-known DPRs are the *Valid*, which randomly selects valid actions, the FIFO that selects the job with the longest waiting time, and the Longest Processing Time (LPT) that selects the job for which the next operation requires the longest processing time. The DPR Avoid Maximum Current Makespan (AMCC) (Mascis and Pacciarelli, 2002) focuses on generating schedules with blocking and no-wait constraints while selecting jobs using a global view (18). DPRs are widely applied and advantageous in environments with a huge and changing variety of products because of their ease of use (15) and moderate computation effort (Gabel, 2009). However, they can only consider one optimization objective at a time (17) and they drastically simplify the JSSP through the selection based on the jobs' properties and are therefore rather imprecise (Gabel, 2009). Context information about the situation in the FMS, i.e., where jobs are being processed and need to go next, transportation dispatching (7), and advanced material flows (11) are not considered, thereby leading to a solution that is not applicable to most real-world FMSs or perform poor.

2.3.2   Reinforcement Learning Scheduling Solutions

To cope with the determined requirements for flexible scheduling systems, we aim to develop a solution that performs near-real-time (3), meaning that it is capable of rapid and reactive decision-making based on the current situation within seconds. We therefore focused on approaches that involve AI, more specifically RL, thereby aiming for a self-learning (15), proactive (4), and flexible scheduling (11–16). Since 1995, RL is presumed to be a promising approach to solve JSSP, when Zhang and Dietterich (1995) were the first researchers to suggest that RL can generate high-performing scheduling using temporal differences to train a neural network to incrementally repair violations of a critical-path schedule. Van Brussel et al. (1998) proposed a very promising approach in which independent learning agents are attached to resources (machines, transportation, etc.), orders (representing a job), and products (holding knowledge about the required processes), and negotiate via a bidding system. This reactive approach fulfills many of the requirements of a flexible scheduling solution (5, 7, 19, 16). However, it involves communication using a protocol and has the drawback of engineering effort for the communication infrastructure, protocol, and rules. Furthermore, agents must learn to interpret the messages for their decisions. Monostori et al. (2006) provided a broad review of various highly sophisticated approaches with societies of agents controlling the whole manufacturing system with different agent groups on different levels, i.e., scheduling agents, transportation agents (7, 9), and assembly agents (Seliger and Kruetzfeldt, 1999). Nevertheless, acceptance of such multi-agent control systems in manufacturing is relatively low because the setup requires high engineering effort (15), scalability and safety can be bottlenecks, and integration into existing systems can be difficult (Monostori et al., 2006). It is thus more feasible to apply RL to certain parts of the manufacturing control system, such as the allocation of resources and transport. Csáji and Monostori (2006) reviewed several RL based resource allocation approaches and concluded that this is a preferable method for use in the industrial domain because RL methods are robust, react to unforeseen (5) and unknown situations (6), can quickly adapt to changes (20), and scale well (12, 13, 14). Nevertheless, there are just a few approaches that can deal with practical flexible JSSPs coping with requirements from the environment of FMSs. In the next section, we analyze these approaches to determine the research gap.

2.4   Identification of the Research Gap

Having a reactive scheduling component in the production control is not yet state-of-the-art in the flexible manufacturing industry and it was shown that offline scheduling approaches are mostly used, whereby a re-calculation needs to be performed when there are deviations from the plan, which is both time-consuming and inflexible. For these reasons, the focus of this work is set on the reactive scheduling component, which includes resource allocation and path planning online during production. The reactive component can be built on the a-priori job prioritization that determines the order of a customer order stack and releases the jobs to the production system accordingly. To give an overview of the most relevant approaches and to determine the research gap, we selected ten representative publications (out of many that were studied) to solve flexible JSSPs in manufacturing from 1998 to 2020. These published concepts were reviewed based on the criteria defined in section 2.2 and summarized in figure 2.6. By rating the alternative research approaches to solve JSSPs we perform step three of the HoQ. Step four involves the identification of the research gap, which can be visually identified by looking at the white spots in figure 2.6. The requirements that are not considered in a minimum of half of the reviewed literature are framed by the rectangle. To determine the research gap, we shaded the top ten requirements that were found by the expert ranking with a shadow. The requirements that are shaded and are located within the rectangle (11–14), represent the requirements that are considered in this research. It can be concluded that most of the considered approaches fulfill requirements (1–6), as the majority use RL with neural function approximators called Deep RL, which has the advantage of quick decision-making during the operation phase because the inference of the trained RL network is rapid. Deep RL solutions can both be proactive and reactive to unforeseen events if the training setup is properly engineered and the training involves various training samples of various and unknown events. It can also be stated that most of the reviewed RL approaches to solve JSSPs considered just a few machines that were located in static positions and complex material flow is respected by very few solutions. Furthermore, transportation times or decisions are mostly completely neglected, in addition to various means of transport. It was observed that if different product variants are considered, the amount of different variants is low, as is the general order volume. Due to this, it was challenging to determine how scalable these solutions are in terms of new product variants. Overlapping skill-sets were considered by almost all approaches, meaning that they can also be used for FJSSPs, where there is flexibility in terms of using multiple machines for each operation. The same holds true for multiple skills per machine, although it was again hard to identify how adaptable the approaches are in regard

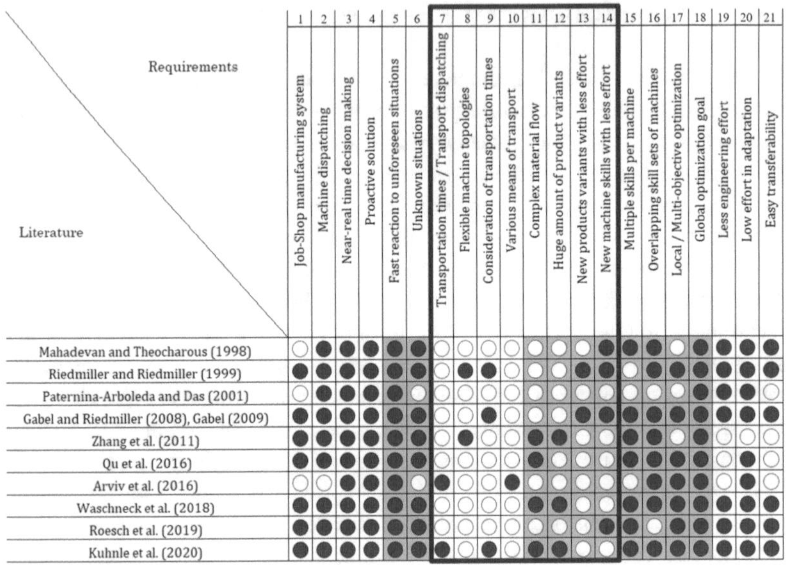

Literature \ Requirements	1 Job-Shop manufacturing system	2 Machine dispatching	3 Near-real time decision making	4 Proactive solution	5 Fast reaction to unforeseen situations	6 Unknown situations	7 Transportation times / Transport dispatching	8 Flexible machine topologies	9 Consideration of transportation times	10 Various means of transport	11 Complex material flow	12 Huge amount of product variants	13 New products variants with less effort	14 New machine skills with less effort	15 Multiple skills per machine	16 Overlapping skill sets of machines	17 Local / Multi-objective optimization	18 Global optimization goal	19 Less engineering effort	20 Low effort in adaptation	21 Easy transferability
Mahadevan and Theocharous (1998)	○	●	●	●	●	●	○	○	○	○	○	○	○	●	●	●	○	●	●	●	●
Riedmiller and Riedmiller (1999)	●	●	●	●	●	●	○	●	●	○	○	○	●	○	●	●	●	●	●	●	●
Paternina-Arboleda and Das (2001)	○	●	●	●	●	○	○	○	○	○	○	●	●	○	●	●	●	●	●	●	○
Gabel and Riedmiller (2008), Gabel (2009)	●	●	●	●	●	●	○	○	○	○	○	●	●	○	●	●	●	●	●	●	●
Zhang et al. (2011)	●	●	●	●	●	●	○	●	○	○	○	●	●	○	●	●	●	●	○	○	○
Qu et al. (2016)	●	●	●	●	●	●	○	○	○	○	●	●	●	○	●	●	●	●	○	○	○
Arviv et al. (2016)	○	●	●	●	●	●	●	○	○	○	○	●	●	●	●	●	●	●	●	●	●
Waschneck et al. (2018)	●	●	●	●	●	●	○	○	○	○	●	●	○	○	●	●	●	●	●	●	●
Roesch et al. (2019)	●	●	●	●	●	●	○	○	●	○	○	●	●	●	○	●	●	●	●	●	●
Kuhnle et al. (2020)	●	●	●	●	●	●	●	○	●	○	●	●	●	●	●	●	●	●	●	●	●

Figure 2.6 Review of approaches for production scheduling in manufacturing systems. The research gap is the intersection of the shadowed top ten requirements ranked by the experts and the rectangle that highlights the requirements that are not fulfilled in at least half of the evaluated literature

to new upcoming machine skills. It can be seen that most approaches optimize their schedules using a global optimization objective, such as the makespan, and that local optimization objectives are also considered. Less evaluation was conducted on the effort in engineering, adaptation, and transferability, which is why these criteria are judged only with respect to the proposed concept design.

To summarize the findings, the white spots in figure 2.6 show the requirements that are not yet addressed by the majority of RL approaches in the literature, but are considered as important requirements from the customer's perspective. Requirements (7–14) are barely considered in other approaches and, in addition, requirements (11–14) are ranked highly by the experts. Many approaches solve very specific real-world JSSP that can hardly be adjusted to highly dynamic and complex production systems. By means of these findings, we defined that our concept should go beyond the current state-of-the-art approaches and should at least consider the requirements (11–14) and the requirements in terms of (5–6) and (15–18). As our goal was to solve FJSSP, we also include requirements (1–2). We use these criteria to define the necessary technical functionalities and to design our concept in chapter 4. To derive the contribution of this work from the identified research gap, the following section concludes this chapter by formalizing the extended FJSSP.

2.5   Contribution of this Work: Extended Flexible Job-Shop Scheduling Problem

Building on the determined requirements, the extended JSSP is formalized in the following section and the contribution of this work is summarized. Zhang et al. (2019) classified JSSPs into 16 different categories according to the characteristics of product demand, number of machines, production environment, type of operations involved, and resource and plant constraints. Zhang et al. (2019) and Lin et al. (2012) re-classified these categories into five main types of JSSPs, namely the basic type, multi-machine type, multi-resource type, multi-plant type, and smart factory type. The basic type involves normal JSSPs that aim to find the optimal operations sequence of jobs that must be performed on specific machines without alternative machines. The FJSSP is more advanced, as jobs can be performed on alternative machines, called Part FJSSP (P-FJSSP). The goal is to determine the optimal sequence of operations, including the selection of machines with respect to minimum makespan and balanced machine workload. Multi-resources FJSSP (MrFJSP) is the enhanced version of FJSSP which is restricted by constraints of manufacturing resources that can occur from set-up times, tools, and transportation, such as conveyors, personnel, or robots. Besides finding the optimal operations sequence and machine selection, the minimum resource transition times are also considered. The extended version of the MrFJSP involves multiple plants and transportation and is known as Multi-plants-based MrFJSPs (MpFJSPs). The fifth category is the MpFJSP with smart factory that places the focus on real-time decision-making and is specified by distributed scheduling entities with self-decision-making.

Our solution is tailored for job-shops or conveyor systems with multiple distributed machines in which manufacturing processes (drilling, milling, grinding), assembly processes (component assembly), and other value-adding or quality-related processes are required (1). It is especially suitable for systems with high dynamics and the possibility for re-scheduling, i.e., for systems where online decision-making is desirable. A rough offline plan can be performed in advance by a solution that prioritizes the orders and determines the release times to the system. Thereafter, our solution finds the optimal sequence of released jobs while considering plant constraints and minimizing the transportation time. While it can be used between multiple work centers or plants, it is mainly developed for dispatching jobs to machines (2) within one work center and for suggesting a switch to another work center if transportation is available. Therefore, it can be categorized in MrFJSPs with the possibility to extend it to solve MpFJSPs in the future. Nevertheless, the MrFJSPs are specified using the determined requirements (IDs) that should be considered in our solution:

- Handling a large number of product variants, as well as new product variants (12–13), independent of the number of operations per job.
- Scaling with the number of machines and skills per machine. We consider machines with overlapping skill-sets (16) that can be extended (14).
- Consideration of a local optimization objective (17), as well as a global objective for all jobs (18).
- Handling of unforeseen events (5), such as machine breakdowns, lacking material or conveyance issues and handling of unknown situations (6), such as new product variants or machine skills.
- Complex material flow should be considered (11), including the transportation times between different machines. Transportation decisions themselves are also included (7).
- It is required to set up a solution that considers low-effort engineering (15).

The following constraints are set to define the setting in which the solution can be applied:

- The prioritization, e.g., by due-dates, and the release of orders are done a-priori.
- Jobs consist of multiple operations, while operations consist of one process step.
- Machines can process one operation at a time
- The set-up times of machines, as well as the loading and unloading of the processed product are added to the processing times and thereby considered.
- The waiting time of a product in a queue depends on the remaining processing time of the current processing of the machine and its queue position.
- The material stock of a machine is not tracked; the material is assumed to be available as long as the material buffer is plugged to the machine.

Reinforcement Learning as an Approach for Flexible Scheduling

After having defined the contribution of this work in chapter 2, the focus is placed on suitable approaches to solve the extended MrFJSP. As in the case of production scheduling, scheduling problems are often a decision-making process of sequences of situations and decisions within a system of complex relations. It was proven to be efficient to distribute the decision making to independent but cooperating entities, such as the drive agents in the warehouse example. Inspired by this, it is investigated how decentralized entities can be advantageously applied to the reactive scheduling component within manufacturing, and therefore, in this chapter we introduce how RL can be leveraged for this. RL is a method of ML to solve sequential decision-making problems by learning entities that interact with an environment by trial and error. The Markov Decision Process (MDP) is the mathematical expression of a problem to be solved with RL and is introduced, together with the RL framework, in section 3.1 and Deep RL in section 3.2. The state of the art in section 3.3 gives an overview of MARL approaches to solve complex tasks with multiple entities.

3.1 Understanding the Foundations: Formalization as a Markov Decision Process

A finite MDP is a formal representation and requirement to solve a sequential decision-making problem with RL. Within MDPs, learners make decisions (actions) based on their current information (states). The overall goal is to find a policy that maps actions to states with respect to a specific goal to maximize the feedback that is received for this action and the future actions. Sutton and Barto (2010) provide a good introduction to RL and MDPs which can be reviewed for a deeper understanding as the following section only gives a brief introduction to the setup in which agents interact with their environment to learn policies.

S. Bär, *Generic Multi-Agent Reinforcement Learning Approach for Flexible Job-Shop Scheduling*, https://doi.org/10.1007/978-3-658-39179-9_3

3.1.1   Agent-Environment Interaction

An *agent* is a learner and decision-maker and interacts with an *environment* with the goal of learning the proper policy to achieve a certain goal. An *action* is chosen from a defined set of actions $A_t \in A(s)$ and propagated to the environment that includes everything but the agent. Agents are called in discrete time steps $t = 0, 1, 2, 3, ...$ and act based on the current *state* $S_t \in S$ that is shown to the agent as the representation of the environment, also known as observation. Figure 3.1 outlines the interaction between the agent and the environment with the overall goal to maximize the reward in the long term (Sutton and Barto, 2010).

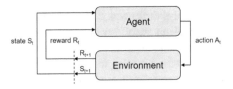

Figure 3.1 Interaction of an RL agent with its environment by Sutton and Barto (2010)

The agent learns how to behave in the environment, first by trial and error, and later by using a policy that estimates the value $q(s, a)$ of each action a in each state s. These state-action values assign credits to individual actions so that the learner is able to select the best action based on the current state while considering the possible effects on future states. An episode is a defined number of steps t in which the agent selects actions in a defined set-up. The sequence of state-action pairs in an episode starts with an *initial state* S_0 and terminates in the *terminal state* S_T, where T is the final step. After one episode, the parameters in the training set-up, the environment, or the policy are adjusted. A subsequent episode is independent of the previous one. Within an episode, the following trajectory is generated and stored for a policy update (Sutton and Barto, 2010):

$$S_0, A_0, R_1, S_1, A_1, R_2, S_2, A_2, R_3, ... \qquad (3.1)$$

The MDP is a tuple $\langle S, A, T, R \rangle$ with a finite set of states S, a finite set of actions A, the transition function T that is defined as $T : S \times A \times S \rightarrow [0, 1]$ and the reward function R defined as $R : S \times A \times S \rightarrow \mathbb{R}$ (Wiering and van Otterlo, 2012). An MDP is *finite* when there is a finite number of elements in the set of states, actions, and rewards (S, A, and R) (Sutton and Barto, 2010).

In a Markovian system, the selection of an action by the agent is solely performed by considering the current state of the system, independently of previous states or actions. Therefore, the current state must consider all necessary information. If the state of the real environment is more detailed than needed for the decision-making, the agent can be shown only a partial observation. With this manually designed partial view, the agent does not need to learn which units of the state are important for the decision-making, as the state only consists of necessary information. If the agent is only able to observe a part of the environment but needs more information for the decision-making, the problem can be described as a Partially Observable Markov Decision Process (POMDP). The characteristics of the Markov property can also be expressed by the probability function giving the same probabilities for the next states every time you visit the state (Wiering and van Otterlo, 2012):

$$P(s_{t+1} \mid s_t, a_t, s_{t-1}, a_{t-1}, ...) = P(s_{t+1} \mid s_t, a_t) = T(s_t, a_t, s_{t+1}). \qquad (3.2)$$

When transitioning from state s to a new state $s' \in S$, there is a probability distribution over the set of possible transitions that is defined by the transition function T : $S \times A \times S \to [0, 1]$. Function T describes the probability of ending up in state s' after executing the chosen action a in state s. This probability function can be used to introduce uncertainties to the system and can be set to 1 if the new state is reached with full certainty. The sum of all probabilities is 1 for each state $\sum_{s' \in S} T(s, a, s') = 1$. For invalid actions, the probability is set to T(s,a,s') = 0 for all states $s' \in S$ (Wiering and van Otterlo, 2012).

A feedback signal from the environment can be used to calculate a scalar reward signal. In the following time step t this numerical *reward* $R_{t+1} \in R \subset \mathbb{R}$ is fed back to the agent to influence the behavior of learning towards the intended goal. As the agent's objective is to maximize the cumulative reward in the long run over an episode instead of just focusing on the single rewards it receives, a good reward function is designed in such a way that the desired goal is achieved when maximizing the rewards over time. The following reward function definition is often used for model-free algorithms, in which the transition and the reward function are unknown, $R : S \times A \times S \to \mathbb{R}$ (Wiering and van Otterlo, 2012). The transition function and the reward function set the boundary of the *dynamic model* of the MDP.

Maximizing the cumulative rewards in the long run formally means maximizing the expected return G_t which, in the simplest case, is defined as the sum of rewards (Sutton and Barto, 2010). By the definition of an MDP, episodic tasks and continuous tasks can be solved. In *episodic tasks*, there is a clear definition of an *initial state*, followed by a sequence of state-action pairs and a defined scenario when the episode terminates, i.e., the finalization of producing a product in the example of scheduling in manufacturing systems, or the selection of an invalid action. The time of termination T can vary and be set to a maximum amount of time steps. *Continuous tasks* can for example be processes that need to be optimized where there is no specific terminate state, e.g., the reverse pendulum example. For such cases, the final time step would be $T = \infty$, thereby leading to an infinite return. This is one reason to introduce the *discounting rate* γ, $0 \leq \gamma \leq 1$ in order to discount the rewards over time. Furthermore, the future rewards can be considered with different weights to facilitate training agents that are more proactive. The agent's goal is now to maximize the expected *discounted return* (Sutton and Barto, 2010):

$$G_t := R_{t+1} + \gamma R_{t+2} + \gamma^2 R_{t+3} + ... + \gamma^{t-1} R_T = \sum_{k=0}^{\infty} \gamma^k R_{t+k+1}, \qquad (3.3)$$

The discount rate determines the weight of future rewards. The reward that is received in future time steps k, is only weighted γ^{k-1} times the actual reward that it would be received immediately. To bind the infinite sequence of returns in continuous tasks, the discount rate should be $\gamma < 1$. For $\gamma = 0$, the agent maximizes only the immediate Reward R_{t+1} and when γ approaches 1, the agent strives to maximize the whole return sequence while strongly accounting for future rewards. To summarize, a trained agent selects actions utilizing a policy and considering possible effects on future state-action pairs, because successive time steps are related to each other by $G_t = R_{t+1} + \gamma G_{t+1}$ (Sutton and Barto, 2010).

3.1.2   Policies and Value Functions

Value function based RL algorithms compute optimal policies by estimating value functions that express *how good* it is for an agent to be in a certain state. The term *how good* stands for the expected return from that state until the end of the episode. When we talk about how to behave in a certain state, value functions are connected with the *policy* that is an ordinary function mapping a state to probabilities of selecting each possible action in that state (Sutton and Barto, 2010). The computational output of a *policy* is an action $a \in A$ for each state s. In *deterministic* cases the policy is defined as $\pi : S \rightarrow A$ and in *stochastic* cases the policy is $\pi : S \times A \rightarrow [0, 1]$. If an agent is following policy π at time t it means that $\pi(a|s)$ is the *stochastic* probability that $A_t = a$ if $S_T = s$. Being in state s using policy π, value $v_\pi(s)$ expresses the expected cumulative return. This is defined in the *state-value* function for policy π (Sutton and Barto, 2010):

$$v_\pi(s) := E_\pi[\sum_{k=0}^{\infty} \gamma^k R_{t+k} | S_t = s], \tag{3.4}$$

where $E_\pi[\cdot]$ denotes the expected value of being in state s following policy π in any time step t. The value of the terminal state is always zero (Sutton and Barto, 2010). How good it is to perform a certain action in a specific state is defined by the *action-value* function $q_\pi(s, a)$ that estimates the expected return of being in state s, choosing action a, and continuing under policy π until the end of the episode (Sutton and Barto, 2010):

$$q_\pi(s, a) := E_\pi[\sum_{k=0}^{\infty} \gamma^k R_{t+k} | S_t = s, A_t = a]. \tag{3.5}$$

The fundamental property of recursive relationships between equation 3.4 and 3.5 is advantageously used in RL and Dynamic Programming (DP). We can thus use the *Bellman equation* for v_π to express the relationship of the value of a state and its successor states (Sutton and Barto, 2010):

$$\begin{aligned} v_\pi(s) &:= E_\pi[\sum_{k=0}^{\infty} \gamma^k R_{t+k} | S_t = s] \\ &= E_\pi[R_{t+1} + \gamma G_{t+1} | S_t = s] \\ &= \sum_a \pi(a|s) \sum_{s',r} p(s', r|s, a)[r + \gamma v_\pi(s')], \, for all s \in S. \end{aligned} \tag{3.6}$$

The expected value of a state is equal to the discounted value to the next states, with the probability of the transition function p, and the following discounted returns until the end of an episode. By this definition, the value information of the following state-action pairs is transferred back to a state from the successor states (Wiering and van Otterlo, 2012) (Sutton and Barto, 2010).

The overall goal is to solve an MDP in a way to achieve the maximum possible rewards over time. Therefore, the optimal policy π_* needs to be found that is better than all other policies π'. More specifically, this means that the expected return is higher or equal for all states $v_{\pi_*}(s) \geq v_{\pi'}(s)$. Even if there are several optimal policies, they all have the same *optimal state-value function*, denoted v_* and defined as $v_*(s) := max_\pi v_\pi(s)$, for all $s \in S$ and they also have the same *optimal action-value* function $q_* := max_\pi q_\pi(s, a)$, for all $s \in S$ and all $a \in A$. Following this, the Bellmann optimality equation for the state-value function can be derived (Sutton and Barto, 2010):

$$v_*(s) := max_a \sum_{s',r} p(s',r|s,a)(r + \gamma v_*(s')).$$ (3.7)

To select the optimal action of the state-value function v_*, the greedy approach of using policy π is applied:

$$\pi_*(s) := argmax_a \sum_{s',r} p(s',r|s,a)(r + \gamma v_*(s')).$$ (3.8)

Coherently, action-value function $q_*(s)$ is defined as:

$$q_*(s) := \sum_{s',r} p(s',r|s,a)(r + \gamma max'_a q_*(s',a')).$$ (3.9)

The relevant action selection can be performed by greedily using the policy:

$$\pi_*(s) := argmax_a q * (s,a).$$ (3.10)

Model-based algorithms are associated with DP, where the value function q is computed using the Bellman update rule to find optimal policies. A perfect model of the environment is a requirement for model-based algorithms. Model-free algorithms are generally linked to RL algorithms, as they do not rely on a perfect model of the MDP but learn by interacting with an environment thereby generating samples of state-action pairs and rewards that are then used to estimate the state-action value function (Wiering and van Otterlo, 2012).

3.1.3   Challenges Arising in Reinforcement Learning

There are several RL algorithms that all have their advantages in certain domains but only work well with a proper training setup. Two major challenges are the credit assignment problem and the exploration-exploitation problem, which are discussed in the following section as they need to be considered when designing the training concept.

The *reward hypothesis* is defined as the maximization of the expected value of cumulative rewards. For some problems, there is only a single reward at the end of the episode, i.e., when winning or losing a chess game or when hitting an obstacle in autonomous driving. In the scheduling domain in manufacturing, a sub-optimal decision to assign a product to a machine can cause delays in the schedule to a later point of time if the agent did not anticipate the needs of a high-priority product within the plant. This leads to the challenge of assigning delayed or sparse rewards to the coherent state-action pairs and is called a *temporal credit assignment problem*. Sparse and delayed rewards make it hard for an agent to learn which of the previous actions led to the failure. In MARLs set-ups, the reward can consist of a local and a global reward share, or of a global reward only that is fed back to the agents, which makes it difficult for the agent to understand the impact of their action on the global goal. Nevertheless, it is possible to learn the policy to achieve the maximum possible reward by using suitable techniques. In Monte Carlo approaches, the reward will be propagated back to the previous state-values at the end of the episode (Sutton and Barto, 2010). When parameterized structures are used to store the policy, the usage of eligibility traces (Harb and Precup, 2017) (Singh and Sutton, 1996), or hindsight experience replay (Andrychowicz et al., 2017) help to update the relevant structures within the memory and therefore to solve the *structural credit assignment problem* (Wiering and van Otterlo, 2012). A further approach is to adjust the estimated value of the state based on the current reward and the value of the next state. This approach is known as temporal difference learning and is the mechanism used in model-free learning. The update rule must not contain the transition or reward function, but rather the reward itself. Algorithms with update rules that perform updates of value estimations after each step are grouped in the class of online RL.

Within the model-free RL algorithms, an important aspect is that agents need to explore the unknown environment by *trial-and-error*. Following a greedy policy would always lead to the action with the highest expected return over the episode. To ensure proper learning, an agent must try out different actions to find a promising action that can be relied on in future. There must be a balance between exploitation of the learned policy to progress the given task, and to receive a good reward and exploring new actions in certain states to allow for improvement of the policy. This trade-off between exploration and exploitation is one of the main challenges in RL (Sutton and Barto, 2010). One strategy can be to control the share of selecting the best action following the greedy policy and exploring new actions in certain states. This strategy is called ϵ-greedy policy with the exploration-exploitation factor ϵ that defines the probability of a randomly selected action. With $(1-\epsilon)$ the greedy-policy is applied to select the action. The hyper-parameter ϵ can be decreased over the course of training to have a high variance in the beginning and less at the end of the training. Another method includes the Boltzmann or softmax strategy, in which the actions are randomly selected but weighted by the relative Q-values, and which is rather applied when using policy-search based algorithms (Wiering and van Otterlo, 2012).

3.2   Deep Q-Learning

In general, there are value function based and policy-searched based RL approaches. As value function based RL methods in combination with function approximation is our first approach in this work, it is introduced in this section although it is out of the scope of this work to give a broad review of all RL algorithms. DP, Monte Carlo (MC) and Temporal Difference (TD) are the main approaches used to learn the optimal value function to derive the policy from it (Bellman, 1957), but only TD algorithms are explored in more detail in the following section. TD is the basis for Q-Learning and Deep-Q-Network (DQN) and has proven to be efficient for problems with chained state-action sequences because predicted effects on future states are reflected in the current state-action value. DQN has successfully been used in (Mnih et al., 2013) and (Mnih et al., 2015) to outperform human-level control in playing Atari games, in (Kempka et al., 2016) to play the first-person shooter game Doom, and in (Gaskett et al., 2000) to control mobile robots using visual data.

3.2.1   Temporal Difference Learning and Q-Learning

TD is one of the most fundamental methods of many RL algorithms in the model-free approach. While interacting with the environment, experience tuples are collected and used to update the estimation values $V(S_t)$ in accordance with the immediate rewards and estimated value of the next step $V(S_{t+1})$. Through this, an agent can adjust its estimations every time it gets new information. Updates along the path are made intermediately and not just at the end of an episode. These updates are only executed on experienced paths and their updates affect the following value estimation (Wiering and van Otterlo, 2012). Any policy π is evaluated by using the following update rule (Sutton and Barto, 2010):

$$V(S_t) := V(S_t) + \alpha(R_{t+1} + \gamma V(S_{t+1}) - V(S_t)), \qquad (3.11)$$

where $\alpha \in [0, 1]$ is the learning rate to control the impact of the value update. This is a hyper-parameter within RL algorithms that can be adjusted during the training to achieve convergence. In TD, raw experiences with tuples of $\langle S_t, A, S_{t+1}, R \rangle$ are used to perform backups every time a transition from one state to the next state is executed. Bootstrapping is used, but only updates the value of the successor state based on the real transition instead of doing the full backup with the weighted average of all possible successor states (Wiering and van Otterlo, 2012). The TD algorithm itself still needs a model for selecting actions based on v_π. Q-Learning, in contrast, learns the Q-functions directly without the need for a transition model and is a proven method in RL in model-free environments (Watkins, 1989) (Watkins and Dayan, 1992). An agent incrementally estimates action values or Q-values for state-action pairs that indicate the *quality* of selecting an action in a certain state. The Q-value function is what an agents strives to learn when using the rewards and the maximum Q-value of the next state by using the following update rule (Wiering and van Otterlo, 2012) (Sutton and Barto, 2010):

$$Q(S_t, A_t) := Q(S_t, A_t) + \alpha(R_{t+1} + \gamma \max_a Q(S_{t+1}, A) - Q(S_t, A_t)). \quad (3.12)$$

The max-operator ensures that for the next state, the Q-value with the highest possible return is used as a characteristic of off-policy learning. Nevertheless, Q-learning is dependent on exploration, and to ensure convergence to an optimal policy regardless of the policy being followed every state-action pair should be visited an infinite number of times with a decreasing learning rate α (Bertsekas and Tsitsiklis ca., 1999) (Watkins and Dayan, 1992).

3.2.2 Deep Q-Network

In the field of value function based RL, the use of DP, i.e., value iteration (Bellman, 1957), has proven to be beneficial to learn the optimal action-value function $Q_i \rightarrow Q^*$ as $i \rightarrow \infty$ whilst storing the Q-function in a table (Mnih et al., 2013). This approach is feasible for a finite number of states, but becomes impractical for advanced problems with a high-dimensional state-space due to computational and storage limitations when using a table. Furthermore, the function is only learned for specific states and is not able to generalize to new states without retraining (Gabel, 2009) (Mnih et al., 2013).

In contrast, neuro-dynamic approaches where Multilayer Perceptron (MLP) neural networks serve as a memory and value function approximation (Bertsekas and Tsitsiklis, 1996) have the great advantage of being capable of generalizing the learned knowledge and applying it to unknown situations, unlike a table that can only be used to look up explicitly known states. This is a suitable and robust technique to efficiently learn complex problems, even if there are no general statements about the convergence behavior of RL algorithms with regard to value function approximation (Gabel, 2009).

TD-gammon was one of the first RL approaches using neural networks to estimate the value function while interacting directly with the environment as an online-learning approach (Tesauro, 1994), with the drawback of having to face some challenges that arise within the so-called online Deep RL. One of the difficulties is that a correlation often exists between the samples in a sequence. This can lead to inefficiency when applying online learning because the learner becomes biased when using the whole sequence for training instead of independent situations. Additionally, experiences can only be used once while interacting with the environment directly, which leads to inefficient data usage. A non-stationary environment can thus become problematic for the learning agent, as the underlying distribution is assumed to be fixed. While following the own policy that changes continuously, the exploration of the environment is directly influenced by the current policy π, which can lead to an unwanted local minimum (Mnih et al., 2013) (Gabel, 2009). The challenges mentioned motivated Mnih et al. (2013) to come up with the more advanced DQN approach—19 years after TD-Gammon—with the difference that DQN is an offline-learning RL method using an experience replay mechanism. Long-Ji Lin (1993) introduced the mechanism of experience replay memory with a "data generation" phase, where a fixed policy such as an ϵ-greedy strategy is used to generate experience tuples from the environment that are stored in the memory. The "value function" phase is executed sub-sequentially with random samples selected out of the pool of experience tuples that are used for updating the Q-function. In this method, the RL agent does not directly interact with the environment but gets samples of unconnected trajectories $\langle S_t, A_t, R_t, S_{t+1} \rangle$ to improve the policy.

DQN leans against Q-learning (Watkins and Dayan, 1992) with the differentiation of using a Neural Network (NN) as a non-linear function approximator to approximate the action-value function $Q(s, a; \theta) \sim Q^*(s, a)$ with θ representing the weights of the Q-network. The model-free RL approach learns the policy through samples from the environment without learning the model of the environment first. During training, it is attempted to minimize the loss function $L_i(\theta_i)$ by changing the proper parameters θ at each iteration i (Mnih et al., 2013):

$$L_i(\theta_i) = E_{s,a \sim \rho(\cdot)}[(y_i - Q(s, a; \theta_i))^2], \tag{3.13}$$

with the target $y_i = E_{s' \sim E}[r + \gamma max_{a'} Q(s', a'; \theta_{i-1}|s, a)]$ for iteration i and $\rho(s, a)$ as the *behavior distribution* over states s and actions a. As the target is dependent on the network weights, θ_{i-1} is fixed while updating the loss function $L_i(\theta_i)$ by differentiating with respect to the weights θ_i (Mnih et al., 2013):

$$\nabla_{\theta_i} L_i(\theta_i) = E_{s,a \sim \rho(\cdot);s'}[(r + \gamma max_{a'} Q(s', a'; \theta_{i-1}) - Q(s, a; \theta_i)) \nabla_{\theta_i} Q(s, a; \theta_i)]. \tag{3.14}$$

A great advantage of DQN is that the experience distribution is averaged over the progress of the policy and thereby helps to avoid the local minimum. Further, the randomized selection of samples from the experience pool breaks correlations and reduces the variance of the updates. As experiences can be reused over several learning phases, data-efficiency is achieved. Especially in real-world applications, it is useful to reuse the sampled data for learning because data is often rare or requires a significant amount of effort to acquire (M. Riedmiller and Dahlkamp, 2007) (Csáji and Monostori, 2008). An additional benefit is that handcrafted feature engineering is no longer needed as this task is fulfilled by the neural network (Mnih et al., 2013) (Gabel, 2009). Impressive results of applying DQN are shown in Mnih et al. (2013), where a sequence of raw images is used as a state representation to outperform human skills in three out of six Atari games. Nevertheless, there is a finite memory size of N that leads to overwriting the buffer with new experiences after N time steps, and random sampling gives equal importance to all transitions. There are thus more sophisticated methods that could be combined with DQN, such as importance sampling (Uchibe and Doya., 2004).

3.2.3 Loss Optimization

In general, there are three different ways to optimize the loss-function with respect to the frequency and amount of training samples: Stochastic Gradient Descent (SGD), batch gradient descent, and mini-batch gradient descent.

In SGD, the loss-function updates are performed using single experiences (Ruder, 2016) (Bottou, 2010). This is advantageous, because the agent does not learn the sequence of states and their correlations, but rather concentrates on single independent situations. With a high update frequency, training with SGD is usually fast, but it has the effect of high variance of training samples that can lead to an oscillating objective function. This poses the risk of jumping out of a local minimum and it is challenging to find the exact optimum. However, when decreasing the learning rate over the training, SGD converges to a local or global minimum for non-convex and convex objective function surfaces respectively (Ruder, 2016).

In batch gradient descent, the gradient is calculated using an entire batch of experiences instead of single experiences and the improved policy is then applied to the environment (Wiering and van Otterlo, 2012) (Mnih et al., 2013). The calculation of the gradient can have a long computation time as the gradients for the whole dataset are calculated to perform just one update, with the learning rate determining the impact of parameter changes. With batch gradient descent, a global minimum for convex error surfaces and a local minimum for non-convex surfaces are achieved (Ruder, 2016). In an alternating batch mode, the steps of data generation by applying the policy in a greedy or exploratory way and the subsequent value function improvements are executed alternatingly (Wiering and van Otterlo, 2012).

Mini-batch gradient descent is probably the most popular way of optimizing the loss-function. With the advantages of batch learning, such as reducing the variance of parameter updates leading to stable learning and being faster in updating, but only using a mini-batch, i.e., with a size of 256 samples, it is used for the gradient calculation. Optimized matrix optimization libraries help to perform efficient computation of the gradients (Ruder, 2016).

Altering batch or mini-batch learning could be beneficial when they are used within a MARL setup because a wide-ranging batch of experiences might be more efficient to train multiple independent learning agents compared to single transition tuples (Gabel, 2009). Having introduced the basics of Deep RL, we analyze the state of the art using multiple agents in the context of RL in the following section, as MARL is a promising approach to solve the MrFJSP as the main contribution of this work. It is further investigated how these decentralized entities can cooperate to achieve the common goal for a complex problem.

3.3   State of the Art: Cooperating Agents to Solve Complex Problems

For straightforward problems that are clearly defined and deterministic, and where no interaction between different components is needed, one single agent is sufficient to cope with this task, such as a closed loop control problem. When focusing on more complex problems that can be distributed to decentralized software components, the use of distributed AI (Weiss, 2013) with multiple entities is beneficial to deal with such problems. Wu et al. (2005) review how distributed multi-agent approaches for resource allocation achieve collaboration of all entities, i.e., by using an interactive market mechanism with negotiation between entities. This approach involves communication using a protocol, with the drawback of high communication costs and a facilitator being needed, which can become a bottleneck. Further research in this field focuses on how to coordinate the learning of strategies of multiple independent agents (Stone and Veloso, 2000). The need for coordination is one of the new aspects that appear when using MARL because (Gabel, 2009):

- The environment is more uncertain, as it is affected by other unpredictable agents.
- Life gets harder for learning agents that strive for convergence, because they have to learn in a non-stationary environment influenced by continuously changing agents.
- Agents have to reason about other agents' future needs together with their possible resulting actions before they make their own decisions. A properly adjusted state input is required to predict other agents' actions.
- The credit assignment problem gains in importance when using a global reward because it is distributed among all agents, even if just one agent has the full liability of the sub-optimal common result. This makes it harder for the agents to learn and understand which actions in which states are good or bad.
- The coordination between the agents is a substantial aspect as this includes stepping back for another agent if this action helps to fulfill the common goal.

There is a variety of research on competitive agents, especially in the game theory. This focuses on agents that act against each other, meaning that every agent has the goal to "win" the game or to get the highest possible return (Filar and Vrieze, 1997). This is an essential approach in problems where there are entities that strive for the same goal but where not every entity can reach this goal (Gabel, 2009). In this thesis, we focus on learning strategies that combine following a local goal and a global goal that is dependent on the common behavior and strategy of all agents. In the MrFJSP is no "winner", because all products must be processed in a certain amount of time with the goal to decrease the overall makespan.

The definition of a Multi-Agent Markov Decision Process (MMDP) is needed because the behavior of the system is influenced by independent acting agents. Gabel (2009) defines the MMDP by a 5-tuple $M = [Ag, S, A, p, r]$ with Ag as the set of m agents, S as the set of states, p as the transition function, and r as the reward function that are both defined over joint actions $(a_1, ..., a_m)$. $A = A_1 \times ... \times A_m$ represent the joint actions of the agents that have an own set of actions A_i in addition to $i \in Ag$.

Because agents conduct distributed learning and do distributed decision-making, the crucial aspect to ensure cooperative behavior is the reward function design. Furthermore, the task of each agent gets extended from making their own decision to predicting the decision of other agents and then selecting their own action based on this additional knowledge, which leads to the need for a proper adjustment of the state representation. The MMDP has the characteristic that all agents are capable of observing the full state of the system and receive the same reward (Gabel, 2009).

3.3.1   Multi-Agent Learning Methods

In general, the two main methods to solve an MMDP using MARL are well explained in Claus and Boutilier (1998), namely Independent Learners (IL) and Joint Model Learning (JAL). In Deep RL, the use of neural networks as function approximators come into play, thereby leading to three additional learning methods: central learning, concurrent learning, and multiple learners that share their parameters by time (Gupta et al., 2017a).

Independent Learners
Using IL in the context of MARL reduces the complexity of the problem to almost that of single-agent learning. The learners are trained while they consider each other only by the non-stationary environment due to the other agents. It is advantageous that single-agent algorithms can easily be applied, however with no convergence guarantee due to the noise of the stochastic environment. Even if this method works well for some problems (Busoniu et al., 2008), the cooperation of agents is basically not given at all, which limits it to problems where not much cooperation is required (Bloembergen et al., 2015) (Claus and Boutilier, 1998).

Joint Action Learners
JAL, however, fully consider other agents in their decision-making. They observe other agents' actions to predict their behavior and learn in the space of joint actions (Bloembergen et al., 2015). In other words, they learn the Q-values of their actions in conjunction with other agents' actions (Claus and Boutilier, 1998). With this approach, agents can perform better than IL in cooperative scenarios but have the drawback that they need to be able to observe other agents' actions. In addition, the complexity increases exponentially with the number of agents because the action space is a combination of the actions of all agents (Bloembergen et al., 2015) (Claus and Boutilier, 1998). Besides the fact that scaling is difficult, there is a disadvantage in being dependent on one decision-maker. In a continuous environment, agents should be able to decide by themselves when a decision is needed, however, with JAL a common action for all agents would be searched at every point in time.

Centralized Learning

In centralized learning, one central controller learns to map a joint model of observations to a joint model of actions, leading to an exponential growth of action and state space with the number of agents. Gupta et al. (2017a) introduce a more practical variant, where the centralized controller is used as independent sub-policies for each agent to choose the individual action based on the joint observation (Gupta et al., 2017a). In systems where homogeneous agents choose discrete actions, the action space is reduced from $|A|^N$ to $N|A|$ for N agents with their action space A. However, the state space still grows exponentially with the number of agents, so this is not yet the most practical way to solve highly complex problems (Gupta et al., 2017a).

Concurrent Learning

Panait and Luke (2005) give a broad overview of the state of the art MARL, the challenges of collaboration between agents, and possible ways to let them communicate either directly or indirectly. They introduce the idea of concurrent learning together with the aspects of credit assignment, the dynamics of learning, and teammate modelling. Concurrent learning is an approach that can be used when a large problem can be decomposed into smaller tasks or areas of improvement. IL can focus on their smaller state spaces to learn a certain task, which significantly decreases the complexity of the search space and computation expenses. This is especially beneficial in scenarios where heterogeneous policies are needed in order to coordinate and fulfill a common goal (Gupta et al., 2017a). With this, we gain much more flexibility in terms of the number of learning entities that are needed to solve the overall task and are therefore again flexible in the computation resources used (Panait and Luke, 2005). Nevertheless, it is more difficult for the single agents, because they still need to take into account that they are not alone and consider other agents' needs in terms of specific actions or states they want to choose. The *dynamics of learning* is what makes it hard for them to learn a stable policy because the co-learners constantly change their behavior and, even worse, they change their behavior by adapting to the original learner's adaptation to them (Panait and Luke, 2005). The *credit assignment problem*—the problem of how to distribute the reward to individual learners—and the *modelling of other agents* are further areas of research in the field of concurrent learning (Panait and Luke, 2005). When concurrent learning can be combined with more advanced techniques to overcome the difficulties mentioned, this method would be suitable for efficiently solving high dimensional problems. In terms of computation effort it would be more efficient if agents shared their experiences although this is only practical for homogeneous agents.

Parameter Sharing

One common issue in MARL is the dynamics of learning, i.e., when multiple agents interact in parallel in an environment, their actions influence the learning of others. The environment results are always non-stationary because of the changing behavior of all agents and their attempts to continuously adapt to the other agents. Maxim (2016) introduced an idea to overcome this problem in Multi-Agent Deep RL. One agent is trained by observing information about other agents, while their policies stay fixed. After a defined period of time, the parameters are shared with the fixed agents and training with the new policies continues. For homogeneous tasks of agents, this sharing of network parameters can accelerate learning and can help to overcome the issue of a non-stationary environment (Maxim, 2016).

3.3.2   Learning in Cooperative Multi-Agent RL Setups

For complex problems, it becomes inefficient when all agents learn in the entire joint state-action space. A good approach to cope with this is to use local observation spaces for agents and consider parts of the local space of other agents as soon as they become interested in the agent's decision. Wiering and van Otterlo (2012) introduced the idea that agents can learn *when* they would like to access the information from their co-agents and when they want to benefit from the advantage of learning in their smaller local action-state space (Wiering and van Otterlo, 2012). Wiering and van Otterlo (2012) writes about the example of intelligent AGVs in a warehouse that have learned to consider other AGV's state views if they are close to each other. The idea behind this sparse level of interaction is to learn as a single-agent when possible in order to be faster and more efficient and to incorporate information from others wherever helpful. The agent would therefore need to learn to recognize states where a broader state view is helpful in addition to the decision-making itself. There are two approaches to learn whether the state of other agents is beneficial to consider. The first, namely *Learning of Coordination*, learns the explicit states in which they need to consider other agents' information using RL and the second, *Coordinating Q − Learning*, does the same based on the observed rewards.

Learning of Coordination

In Learning of Coordination, Spaan and Melo introduceInteraction-Driven Markov Games (IDMG) ((Spaan and Melo, 2008) that can be solved by an algorithm presented by Melo and Veloso, which can learn in which state additional local information is needed without handcrafting it beforehand (Melo and Veloso, 2009). The action space of each agent is enhanced by a pseudo-coordination action (*coordinate*) that is used to perform an active perception step when needed. When an agent selects this action, information, e.g., about the location of other agents, is provided to the agent, and this action will later be used to decide if additional information should be obtained. The reward is designed in such a way that using the *coordinate* action unnecessarily is not punished as severely as not coordinating. This makes the agent learn the interaction states of the underlying IDMG (Wiering and van Otterlo, 2012).

Coordinating Q-Learning
In Coordinating Q-Learning (CQ-Learning) the algorithm learns three tasks: identifying conflicting situations in which information about other agents is needed, selecting the action, and updating the Q-function. Agents learn about conflicting situations by using statistical comparisons. The agent is thus first trained in a single-agent environment to have a baseline, and compares the received rewards for the state-actions pairs with the rewards received in the multi-agent environment. When the agent experiences deviations for the same state-action pair, it will learn that this situation is influenced by other agents and that special care is needed here (Wiering and van Otterlo, 2012). Before selecting the action in such cases, the algorithm inspects the state information of the other agents and verifies if the joint state is the same as it was in the conflicting situation identified. If so, the agent will select the action based on the joint state information, or otherwise on the local state information. When the current state is not marked as a special case, the action selection is done as if the agent acts in a single-agent environment (Wiering and van Otterlo, 2012). The Q-function is updated by bootstrapping the Q-values of the states that were complemented. From the information about conflicting states, the coordination dependencies between the agents can be derived (Y-M. De Hauwere and Nowé, 2010). This information can be useful and transferred to a more complex environment (Vrancx, 2011) where agents can focus on their explicit task while using the transferred knowledge of coordination (Wiering and van Otterlo, 2012). YM De Hauwere (2011) adjusted this approach by observing the reward signal in subsequent steps to detect sparse interaction, such as identifying the relevance of the order of arriving goods in a warehouse (Wiering and van Otterlo, 2012).

   Another approach is to learn joint action-values only where needed and use single actions elsewhere. It is problematic that these joint action-values are only needed in very specific situations which can be described in Coordination Graphs (CGs) that represent the dependencies between agents (Guestrin, 2002b) (Guestrin, 2002a) (Kok, 2004) (Kok, 2006). This is known as Sparse Cooperative Q-Learning (SCQ). The design of CGs can be very complex with a high engineering effort, and it is thus hard to realize for complex problems. This is why Kok (2005) presented Utile

Coordination where these CGs are computed automatically by statistical information about dependencies of other agents when receiving rewards. However, this approach needs the entire state-space to select the actions that lead to an exponential growth of complexity with the number of agents (Wiering and van Otterlo, 2012).

In general, MARL is a well-studied field to solve MMDP (Panait and Luke, 2005) (Busoniu et al., 2008). Tan (1993), Panait and Luke (2005), Bloembergen et al. (2015), and Gupta et al. (2017a) provide comprehensive studies in the area of cooperation between interacting agents in MARL setups. Tan (1993) compared the performance of independent agents to three ways of cooperation in MARL systems: communicating information instantly, making use of episodic experience, and sharing knowledge regarding epochs. He found that the first cooperation method, namely to share information directly, is the most promising one to learn and converge faster when solving a cooperative task (Tan, 1993). How crucial a proper reward design is for achieving cooperatively acting agents is demonstrated in Tampuu et al. (2017), where he used a multi-agent extension of DQN to demonstrate competitive and collaborative behavior, dependent on the reward schemes, while playing Pong. Bloembergen et al. (2015) combined the dynamics of evolutionary game theory and RL and provided an overview of advances in the study of evolutionary dynamics of MARL. This analytical study offers inputs for gaining insights in the learning dynamics and to better compare the behavior of different RL algorithms and approaches (Bloembergen et al., 2015).

3.4 Summary

In this chapter, the fundamentals of RL and the formalization of the problem as MDP are introduced. The overall goal is to influence the behavior of RL agents in their environment to achieve a policy that maximizes the cumulative reward to solve a problem. Q-Learning is a fundamental method for many RL algorithms and has proven to be successful when combined with neural network structures. DQN utilizes Q-Leaning and neural networks to efficiently solve problems with chained state-action sequences. One challenge that arises is the credit assignment problem, which gains even more importance when multiple agents aim for a common goal and receive a global reward. This feedback is propagated to all agents, even if just one agent is responsible for the sub-optimal or optimal outcome. Therefore, cooperating agents are trained with MARL approaches with the challenge to learn in a non-stationary environment affected by the changing policy of all agents. Agents have to predict other agents' actions in order to coordinate and choose sub-optimal short-term actions whenever it is helpful for the long-term team goal. Approaches to train cooperating agents in the MMDP are central learning, concurrent learning, and parameter sharing, whereas combining them with suitable techniques for coordination can lead to an efficient training strategy that is meant to be found.

Concept for Multi-Resources Flexible Job-Shop Scheduling

4

After having introduced the scheduling problem together with the state-of-the-art approaches used to solve it, we found that many of these approaches lack in terms of meeting the requirements that we defined for the MrFJSP in section 2.5. Therefore, in step six the HoQ proposes to translate these requirements into technical functionalities and to evaluate the dependencies and relations between both sides in steps seven and eight. We consequently introduce our concept of an agent-based scheduling approach considering these technical functionalities. The RQ1 is thereby addressed in this chapter. After introducing the overall concept in section 4.1, we formalize the problem as MDP with possible action, state, and reward designs in section 4.2. The concept of utilizing agents to control carts and the training concept using a Petri net were already published in Baer et al. (2019) and Baer et al. (2020a). The state design and reward concept were published by Pol et al. (2021) and Pol (2020). Section 4.4 compares the requirements of the market view with the implemented functionalities of the concept before the concept is evaluated in chapter 5. As the generic concept is applied to an example FMS for the evaluation, we introduce the exemplary FMS including the digital representation in section 4.3.

4.1 Concept for Agent-based Scheduling in FMS

The RQ1 addresses the formalization of the MrFJSP to fulfill the requirements of reactive production scheduling that were specified in section 2.5. The overall goal is to develop an online scheduling system that fulfills the requirements of being flexible in terms of overlapping machine skills and a changing environment, as well as being scalable to a high number of product variants and robust to unforeseen events. In the following subsections, we introduce the overall concept for production control and the use of job specifications and Petri nets for training purposes.

S. Bär, *Generic Multi-Agent Reinforcement Learning Approach for Flexible Job-Shop Scheduling*, https://doi.org/10.1007/978-3-658-39179-9_4

4.1.1   Overall Concept

The concept is to train RL agents that control products through the FMS, including routing and resource dispatching. The decision was made based on the analysis of research about RL-based scheduling solutions in section 2.3 and the identification of the existing research gap in section 2.4, which showed the research potential of RL approaches used for reactive scheduling. Our concept is that one agent instance controls one cart, on which a work-piece carrier and a certain product are located and assigned until the production of the product is finished. With the assignment of agents to products, we have the advantage that the transport can be controlled (7) in addition to the resource dispatching, which is in contrast to most other approaches where the agents are directly assigned to the manufacturing module or machines, as seen in the work of Gabel (2009). One further advantage is that when the agent follows a plan to reach a certain machine, the agent can change this plan on its way to the machine, as the decision-making is done step-wise on each decision-making point. The hypothesis is that this design approach makes the solution very flexible and robust to unexpected situations (5) and by the use of deep NNs in combination with RL agents, we assume that the agents can generalize to unknown events (6).

4.1.2   Job Specification

When a product is configured using a product configurator, which is one component of the *Product Lifecycle Management System*, the *bill of processes* and *bill of materials* are derived and transferred to the MES. The bill of processes lists the processes needed and the bill of materials lists the materials needed to manufacture the product. The processing of the data in the MES is provided in section 7.2.1 in detail, but it can be preempted that the available resources (e.g., a machine with available materials and processes, and available transportation to the machine) are filtered by the required resources (processes and materials) to manufacture the product to derive the *job specification*. The job specification describes how a product can be manufactured in the current setting. It lists all sets of $[M, T_M]$ for each machine M that can process the relevant operation of a job with the expected processing times T_M. An example of job specification is given in table 4.1, which shows one job with four operations and two alternatives each, where the first operation can be processed by machine one in one unit of time or by machine five in eight units of time and so on. The operations must be executed in the given order.

Table 4.1 Structured representation of a job specification with four operations and two alternative machines per operation with according units of time for the considered job specification: [1,1]v[5,8]; [2,4]v[4,5]; [6,8]v[1,3]; [3,7]v[5,1]

Operation	Machine	Time	Machine	Time
1	1	1	5	8
2	2	4	4	5
3	6	8	1	3
4	3	7	5	1

4.1.3   Petri Net Simulation

The design of a suitable digital training environment is accomplished by considering the criteria of computational efficiency, engineering effort for set-up and adaptation, and costs, weighted in the order mentioned. For the considered example FMS used for the evaluation, the straight-forward alternative would have been to use the existing digital twin of the FMS that was implemented with Siemens Tecnomatix Plant Simulation. With this simulation software, all necessary properties, such as transportation times, can be depicted. The major drawback was the execution time of one simulation run of approximately five seconds. The engineering needed to be done in the tool itself and the usability to connect it to the Python training environment to create various different training scenarios was not given without bigger effort. Further, to develop a concept that depends on this software tool is costly due to the license fees. The goal is to develop a concept that can be generally applied and set-up for different FMSs easily. For these reasons, we looked for alternatives and found that Petri nets are used in the domain of FMS for simulation and can be applied by anyone without any license fees or specific software because it is a modelling concept (Zhou and Venkatesh, 1999). The advantage of using a Petri net is the ease of depicting a complex FMS, the simple implementation, and the adaptation with less effort. For any plant design, the Petri net and the related incidence matrix can automatically be derived from a rough depiction of the plant design, which makes the set-up easy. The same applies for changes to the plant design, which can be considered by adjusting the Petri net accordingly. Furthermore, the mathematical representation of a Petri net is very advantageous in terms of computation efficiency, which will be explained in more detail.

As defined in Baer et al. (2020a), we use a Petri net consisting of p places that represent p locations where a product can be located and tr transitions to move a token between these places, which represents a product in the real plant.

We discretized the locations in the FMS into the p places, and when a product is located between two places, it is considered to be in the next place for the next decision-making. The self-transitions let the product stay in the same machine for the subsequent operation. For FMSs with transportation times of various lengths, the times must be modeled in the Petri net simulation.

The state of a Petri net is defined by one specific token representing a certain product, which can be moved between the places. As a demonstration, figure 4.1 depicts the Petri net of our example FMS which is further introduced in section 4.3. If one token was located in place 7, transition 1 and 2 would be activated and could be fired by the agent to move the token from place 7 to place 1 or 8 respectively. Places 1–6 represent the decision-making points on the conveyor belt where the agent has

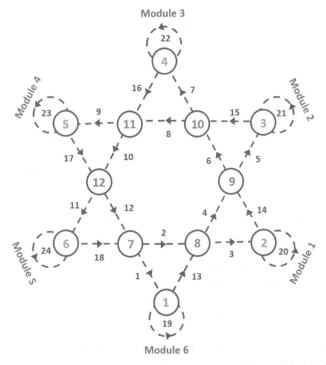

Figure 4.1 Petri net representation of an example FMS with places 1–6 as decision-making points within the machines and places 7–12 on the conveyor belt. The transitions 1–18 move products from one place to the next and 19–24 are self-transitions to stay in a machine

to decide whether to stay on the conveyor belt or to direct to a machine. Places 7–12 depict decision-making points at the machine interfaces. On these places, the agent is asked to decide either to stay in the machine that is connected to this interface for the subsequent operation, or to transition to the next decision-making point on the conveyor belt. In the real system, transitions can be any means of transport, e.g., a conveyor belt, a handling system, or an AGV. When considering AGVs, the estimated transportation times between places must be adjusted accordingly. Transitions indicate that a certain place can be reached from the current place.

By using the incidence matrix C, the plant structure and possible transitions at each place are encoded with much less engineering effort (15). When a transition is selected for job j_1 at the current place $L_{j_1 1}$ that is represented by the transition vector t_1, the new place $L_{j_1 2}$ is obtained using the following equation:

$$L_{j_1 2} = L_{j_1 1} + t_1 \cdot C. \tag{4.1}$$

The addition of the dot product of the transition vector t_1 and the incidence matrix C to the current state marking of the considered product $L_{j_1 1}$, results in the new state marking $L_{j_1 2}$ and therefore the new place of the product. The vectors $L_{j_1 1}$, $L_{j_1 2}$ and t_1 are one-hot encoded, meaning that only the entry that is 1 represents the marking of the current place and transition respectively—all other entries are 0. When there is more than one product within the FMS, equation (4.1) has to be conducted every time a product reaches a decision-making point. With the modeling of the FMS using the incidence matrix C, very little computation effort is needed to calculate the new location of a product.

One agent would follow the places and transitions in the Petri net in figure 4.1 and would be called on each place when no waiting time is assigned to that place (waiting times are assigned when there are other products in front in the queue of a machine or because of the processing time of the product within a machine). For the assigned waiting times, the agent is just called when the waiting time is over. According to this, agents are called in time intervals with lengths that are determined by the waiting times or processing times of machines, similar to the concept of Gabel and Riedmiller (2008).

4.2   Formalization as a Markov Decision Process

We formalize the MrFJSP as MDP following the concept published by Baer et al. (2020b) and Pol et al. (2021) and therefore present the action, state, and reward designs used in the proposed approaches. We define that a full observation of the current state is possible at any time, but we select a relevant partial observation for efficient learning. The considered MrFJSP is deterministic, as long as only one product is manufactured at the same time, meaning that a certain action a in a given state s will always result in the same following state s', and receive a reward r. In the simulation, we do not include hardware-related faults that can, for example, result from jammed switches because it is not the goal to learn and predict such incidences, but rather to react on them when an incident happens, as evaluated in section 7.3.

4.2.1 Action Designs

The task of an agent is to navigate their product through the plant, including the assignment to the preferred machines. Translating this into the simulation setup, the action of one agent is to move their own token that represents the controlled product within the Petri net from one place to the next, which includes the dispatching to the relevant machines when a token is moved to a machine interface place where a machine is connected to the system.

We propose two possible action designs. The action space of an agent can be defined as the number of all possible transitions A_{all}. Alternatively, the maximum number of transitions of all decision-making points is determined and is defined as the action space A_{small}. In the exemplary Petri net of figure 4.1, the maximum number of transitions for all places A_{small} is two.

Gabel (2009) considers changing action sets where machines are defined as agents that select certain operations out of a variable set of operations. He points out that with this design, actions can be executed multiple times so that recirculating tasks can be modelled easily. With our concept, the action space is independent of the number of jobs and operations per job. Furthermore, we can deal with the recirculation of products in the FMS because our concept is able to control products including transportation decisions, which naturally allows products to recirculate in the FMS (11). In addition, we are independent of the number of possible transitions in each place with the action space variant of A_{all} actions, meaning that on some decision-making points there could be three, four, or five possible transitions. When a new machine is introduced to the system (14) the action space is not affected as it is only dependent on the transport options to the machine interfaces. If a new way of transport is introduced, such as AGVs, both action space variants need to be adjusted and the agents need to be re-trained. The Petri net also needs to be extended by new transitions between the places, and the action space variants need to be adjusted by the additional options.

Table 4.2 Comparison of the two action spaces

	A_{all}	A_{small}
Independent of jobs and operations	x	x
Situation-based decision making	x	x
Re-circulation of products in the FMS	x	x
No adjustments for new machines	x	x
No adjustments for new transport		
Flexible number of transitions per place	x	
Only valid actions can be chosen		x

Our two action space variants consist of A_{all} and A_{small} actions, referring to all possible routing directions (Baer et al., 2020b) and the maximum number of routing directions at each place. The characteristics are summed up in table 4.2. With the concept of calling the agents at the discrete decision-making points, even if their next destination is planned, agents have the advantage of being able to change the plan on their way. This is beneficial when a more efficient means of transport appears, because an AGV approaches the agent's current location when a highly prioritized product crosses the agent's way, or in the case of a machine breakdown (5). With this situation-based decision-making, in each situation the decision is made based on the latest information propagated to the agents by the state.

4.2.2   State Designs

We designed two state variants that are later evaluated in the experiments. The naive state vector is used to find a suitable RL design and training approach and the advanced state design is subsequently used to fine-tune the agents' behavior. The state is designed considering the defined requirements, referenced in brackets. The state should enable the agents to react to machine breakdowns (5), to consider transportation times (7), and various machine topologies with machines with multiple or overlapping and new skill-sets (19, 16, 14). The hypothesis for the advanced state design is that it also enables the agents to deal with a large number of product variants (12) and new variants (13), to cooperate to achieve a global optimization objective (18), and to generalize their knowledge to new situations (6). We designed our state vectors by selecting information that we assume would help an agent to interpret the current situation in the FMS, including the needs of other agents and information that relates to the expected costs in the future. The information used should be computable from the available data from the FMS.

Naive State Design

For the naive state design, we follow the state design of Maxim (2016), as it was previously shown to be successful and enables independent agents to solve a complex problem. In general, it considers information about the agent taking the decision, information about the background, and information about the other agents. Figure 4.2 visualizes the components of the naive state representation that consists of the current location of all products, the progress of the controlled product in processing the job, and information about the current machine topology (Baer et al., 2020b). With this state design, the agents are able to learn the following context:

- current location of the agent itself and other agents
- distances to the machines depending on the agent's position
- machine availability in each position, including machine breakdowns
- queue status of all machines (implicit by location of agents)
- whether a machine is occupied (implicit by location of agents)

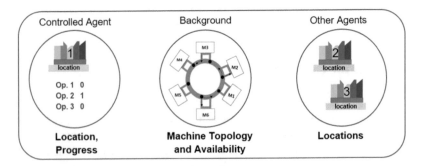

Figure 4.2 Components of the state representation with information about the controlled agent and other agents. Information about the background includes the machine topology and availability, with the example order (1, 2, 3, 4, 5, 6)

This state design does not include information about the available machines for each operation of the controlled product. The purpose of using this state design is to find a good training set-up and strategy, and therefore it is kept lean. The job to be fulfilled by the agents is kept static for these experiments and varies in the experiments using the advanced state design that includes the information about the job specifications.

Similar to Maxim (2016), we use a one-hot encoded vector that should make it easier for the agent to learn different information. The information is concatenated in a uniform input vector, as the information could have huge differences in their values ranges when they are fed to the agent directly (e.g., 1 for the progress and 512 for the topology number). The first component is a one-hot encoded vector representing the current location of the controlled agent. The length depends on the total number of positions in the Petri net. The 1 indicates at which of the p places the agent is currently located.

We also use a matrix to replicate the locations of all agents in the FMS, in addition to information about their queue position and their ID in the first, second, and third axis respectively. The size depends on the queue length of each place and on the maximum number of agents in the FMS. The progress of the agents can be described by a one-hot encoded vector with the size of the maximum number of operations, which we limited to four in the naive state design, while unlimiting it in the advanced state design. The 1 indicates the next operation to be processed. Lastly, we include the machine topology, which defines the order of all available and unavailable machines around the manufacturing plant, as a matrix with the columns representing the machines interface places where a machines can be connected to the FMS. The rows represent the manufacturing modules that are docked to the certain machines interfaces. Failures are treated in the same way as if no module is docked at this position and are encoded by a 1 in an additional row.

We concatenate the one-hot encoded components and flattened the resulting matrix to shape the naive state input vector so that we could feed it to the agent at each decision-making point.

Advanced State Design

To solve the requirements of the defined MrFJSP, it is necessary to handle various jobs' specifications and schedule them optimally instead of being specialized for a single product. We therefore enhance the naive state design by an advanced encoding of the job specifications and include it in the state representation that can be seen in figure 4.3. Therefore, we can neglect the progress vector as this get lapsed with the advanced job encoding. With this, the agent conducts the decision-making based on their own job specification, as well as the job specification of other agents. This information completes the state input that is necessary for being able to perform proactive decision-making. With the advanced state design, the agents are enhanced by the following information:

- required machines of the controlled agent and other agents for the current and for the next n steps with the predicted processing times
- whether a machine is occupied and when it will be available (implicit by agent location matrix and job specification of other agents)

Our hypothesis is that the advanced state design is the basis for cooperation between agents and further supports the development of a scalable solution that generalizes to different job specifications. Especially for the generalization aspect, the proposed encoding is very beneficial as it forces the agent to interpret the job specification, which is the requirement for being able to react to unseen job specifications of new product variants. This is a crucial aspect, as it would take unreasonably long to show the agent all kinds of job specifications during training and there would be no certainty that new job specifications can be handled or whether the agents only learn the shown job specifications. Practically, jobs that result from orders in an FMS can be a variable in the number of operations and the possible machines per operation. Variable sizes of the job specifications cannot be included in the state input, as the state input of Deep RL must have a fixed size. To address this issue, we introduce the job encoding illustrated in figure 4.4, where T_{o_i,m_m} is the time it takes for machine m_m to process operation o_i. The advanced job encoding allows the job specification to have any number of operations by introducing a lookahead. This means that the agent only sees a small section of the entire job specifications comprising the next n operations.

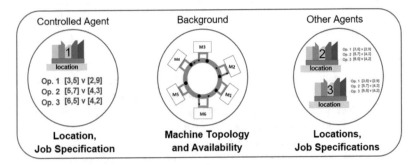

Figure 4.3 Enhanced state representation with additional information about the job specifications of all agents, their location, and the machine topology

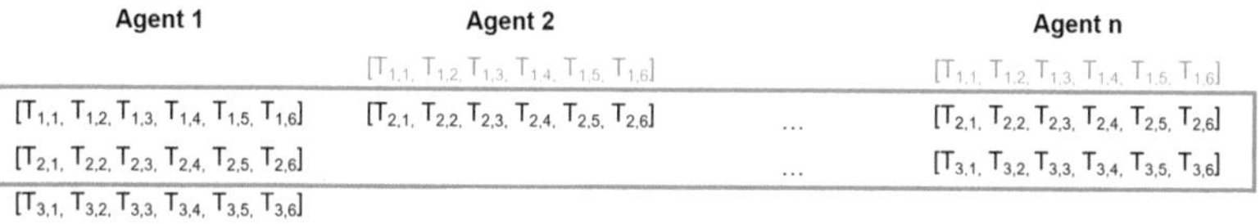

Agent 1	Agent 2		Agent n
	$[T_{1,1}, T_{1,2}, T_{1,3}, T_{1,4}, T_{1,5}, T_{1,6}]$		$[T_{1,1}, T_{1,2}, T_{1,3}, T_{1,4}, T_{1,5}, T_{1,6}]$
$[T_{1,1}, T_{1,2}, T_{1,3}, T_{1,4}, T_{1,5}, T_{1,6}]$	$[T_{2,1}, T_{2,2}, T_{2,3}, T_{2,4}, T_{2,5}, T_{2,6}]$	...	$[T_{2,1}, T_{2,2}, T_{2,3}, T_{2,4}, T_{2,5}, T_{2,6}]$
$[T_{2,1}, T_{2,2}, T_{2,3}, T_{2,4}, T_{2,5}, T_{2,6}]$		...	$[T_{3,1}, T_{3,2}, T_{3,3}, T_{3,4}, T_{3,5}, T_{3,6}]$
$[T_{3,1}, T_{3,2}, T_{3,3}, T_{3,4}, T_{3,5}, T_{3,6}]$			

Figure 4.4 The advanced job encoding allows the inclusion of the job specifications in the agent's state using the lookahead n for the current and next n operations

The lookahead is set to two and is indicated by the gray rectangle, but is a hyperparameter that can be adjusted. The job encoding considers the job specification of other agents with the same lookahead. Depending on the progress of other agents, an agent can see what their next n operations are and which machines they need, which may be useful for cooperative behavior. To address the issue of considering a variable number of possible machines for each operation (16), the advanced job encoding considers every machine in the manufacturing system. The size of the vector is set by the maximum number of machines within the FMS, which we defined to be six in our example representation. Therefore, the nth operation of a job specification has six values $T_{n,1}, ..., T_{n,6}$, stating the time each machine requires to process the operation. The processing time of each machine should be normalized and discretized.

If an agent has finished the job, or is removed from the environment due to making a mistake such as choosing an incorrect transition or machine, the corresponding section in the state vector is padded with zeros. The same applies if there are less operations remaining than the size of the lookahead. It should be noted that the advanced job specification encoding does not provide a solution for a variable number of agents by default. With a variable number of agents in the system, it is possible to consider a subset of the most relevant agents, the agent itself, and the n others selected by some eligibility criterion. For applying the concept using the example FMS that is introduced in section 4.3, we used a fixed number of three agents, with each agent able to see the lookahead for its own job specification and that of the two other agents.

With the additional information, the agents should be able to deal with a large number of product variants (12) and unseen product variants (13). By integrating the job specification into the advanced state design, overlapping skills of the machines (16) and machines with multiple skills (19) are included. In chapter 6 it is further evaluated whether the advanced state design provides the requirements for cooperation between agents and for generalization to unknown situations.

4.2.3   Reward Design

Our overall goal is to train agents that fulfill their own job with respect to their individual optimization objective (17) and at the same time consider a global objective (18). We define the processing time of the controlled product as the local optimization goal, but the reward design is developed so that it is independent of the considered objective. The first goal is to make the agents learn to successfully fulfill their job and then to optimize their policy considering the local objective. We aim to achieve this by using the local reward function during training. The ultimate goal is to train agents that consider their local objectives in addition to the global objective, which forces the agents to cooperate because it can cause a conflict with their own goal. We also aim to design a global reward function that encourages cooperative behavior to fulfill the common objective that is defined as the minimum makespan of all products within the considered order stack. As the optimization objectives are represented as normalized values in a certain range, it is straightforward to transfer the reward concept to other optimization objectives. We present the concept of our local reward design together with the different global reward concepts to enable cooperative agent behavior that is evaluated in chapter 5.

Local Reward Design

We use a dense local reward, meaning that the reward for each agent is calculated after every step within an episode based on the individual performance with respect to their optimization objective. The agent receives the reward only when a decision was made. Therefore, the agent receives no reward while waiting in a machine during processing or while waiting in the queue in front of a machine, because the waiting time is assigned to the reward for reaching this state. The objective is that the agents learn to choose the fastest option to process their operations, including choosing the faster alternatives when there are already other products waiting in the queue in front of a machine, if there is an available alternative-always with respect to coherent transportation times. We therefore use the processing time of the individual job to calculate the local reward after every step. We give each transportation step a slightly negative reward $R_{transport}$ to encourage the agent to limit the number of steps on the conveyor belt to as few steps as is necessary, in parallel to finding an optimal schedule. From our experience, we found that it is necessary to reward the agent with a positive compensation reward R_{comp} whenever it has reached a machine that can process the current operation, so that the agent does not lose the entire reward on the way to the machine through the transportation steps. This compensation reward must be selected so that it can compensate the maximum traveled distance to reach a machine that is needed. Additionally, we add the inverted processing time to motivate the agent to prefer machines that are faster in processing the operation. If the selected action led to an invalid transition, the reward $R_{invalidsteps}$ for this action is received and the agent is removed from the simulation. We equally treat the transitions to a machine that cannot process the current operation or to unavailable machines, regardless of whether they be un-docked, lack material, or have any kind of failure as shown in Baer et al. (2020b). For these actions, the local reward is defined by $R_{invalidmachines}$.

The local reward design $R_{local}(s, a, s')$ is described by the following equation:

$$R_{local}(s,a,s') = \begin{cases} R_{transport} & \text{valid transportation steps} \\ R_{comp} + 0.1 \times R_{maxprocessing} + (-0.1 \times (T + W)) & \text{valid machines} \\ R_{inavlidsteps} & \text{invalid steps} \\ R_{invalidmachines} & \text{invalid machines} \end{cases} \quad (4.2)$$

where T is the time required to process an operation on the selected machine (Pol et al., 2021). T is the normalized processing time that ranges from 1 to 9, as defined for the job specification. We need to add the maximum processing time $R_{maxprocessing}$ to compensate for adding the inverted processing time. However, the reward can exceed this range when there are other agents waiting in the queue of a machine, because in these cases the inverted waiting time W is added to T. The hyper-parameters can be selected depending on the properties of the FMS, but we present default parameters in section 5.2 that were found by parameter-studies for our example FMS the reward design was applied to.

Global Reward Designs

Besides the local reward that has the goal to train agents to fulfill their job following a local objective, a global reward component is required, which is distributed equally among the agents to enable cooperating behavior. We therefore present a dense global reward design, which is a combined reward of the dense local reward component and a global reward factor. The dense reward provides specific feedback to the agents for every action, but requires a proper reward engineering. We also present a sparse reward design, which feeds back only one reward to all agents at the end of an episode. The sparse reward design is straight forward in the engineering, but can lead to the credit-assignment problem, as agents do not know explicitly how high their contribution to the reward was.

For the dense global reward design, we need to calculate the global reward factor that should adjust the local dense rewards. The total reward therefore either becomes slightly larger in the event of a good total makespan, or slightly smaller in the event of a bad total makespan to fine-tune the agents' behavior with respect to the global optimization goal (Pol et al., 2021). As the global optimization objective is defined to be the total makespan that is a common optimization goal for the JSSP, the calculated global reward factor can only be applied when all agents finish their jobs within the considered episode, so that the total makespan is comparable. To judge whether the total makespan is good in comparison to the optimal schedule, an estimation of the optimal makespan is used as the makespan cannot be calculated ad hoc for unknown cases (Pol et al., 2021). To estimate the optimal makespan, we calculate an upper bound b_U and lower bound b_L, in which the makespan is likely to be found for the considered jobs. Figure 4.5 shows the scheme that is used to calculate the lower bounds for each job specification individually, exemplarily shown for three job specifications js_1, js_2, and js_3. The calculation of the upper bound is performed equally by summing up the processing times of the slowest machines in addition to the transportation times between those machines (Pol et al., 2021).

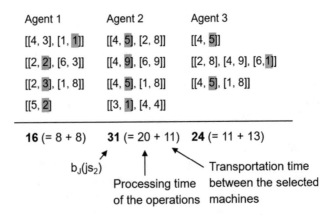

Figure 4.5 Calculation of the lower bound for three given job specifications (Pol et al., 2021)

As our training approach is to randomly determine combinations of job specifications at the beginning of each epoch as is further described in chapter 5, it is unfeasible to calculate the bounds of the considered problem instances beforehand. We therefore need to calculate the bounds at the start of each epoch. We calculate the global bounds of the considered job specifications js_i for agent i by determining the maximum of the individual bounds:

$$b_{LG} = max(b_L(js_1), b_L(js_2), b_L(js_3)) * w_1 + w_2$$
$$b_{UG} = max(b_U(js_1), b_U(js_2), b_U(js_3)) * w_3 + w_4$$
$$(4.3)$$

with b_{LG} and b_{UG} as the global lower and upper bounds determined by the maximum local lower and upper bounds of each job b_L and b_U respectively (Pol et al., 2021). The global bounds are selected by the max operator of the individual bounds because the makespan is defined as the processing time when the last operation of a certain set of jobs has finished. In the case where all agents finish within their upper bounds, the total makespan is defined by the maximum of the upper bounds (Pol et al., 2021). It should be noted that these bounds are rough estimations and do not consider the timings of different products using the same machines that go along with possible waiting times in the queues of a machine, which are neglected. The empirical constants $w_1, ..., w_4$ are introduced to adjust the bounds to address this problem.

Using the determined lower and upper bounds, we empirically define a function with which the global reward factor $R_{factor}(c, f_1, f_2, f_3, f_4)$ can be calculated using the empirical constants $f_1, ..., f_4$:

$$R_{factor} = \begin{cases} f_1^{\frac{m-c}{m-b_{LG}}} & \text{if } c \leq m \\ -\frac{f_2}{b_{UG}-m}c + \frac{f_2 m}{b_{UG}-m} + f_3 & \text{if } m < c \leq b_{UG} \\ f_4 & \text{if } c > b_{UG} \end{cases} \qquad (4.4)$$

with b_{UG} and b_{LG} as the global upper and lower bounds for the job specifications, m as the mean value, and c as the total makespan (Pol et al., 2021).

The dense local reward of the individual agents is multiplied with the global reward factor to increase the local reward for a smaller makespan than the mean and decreases the total reward for a makespan bigger than the mean (Pol et al., 2021). We exceptionally treat negative local rewards, which we do not boost to avoid discouragement of exploration (Pol, 2020). The dense global reward $R_{globaldense}$, received after every action of an agent, is thereby defined by the following equation:

$$R_{globaldense} = R_{local} \times R_{factor} \quad \text{if } R_{local} > 0 \qquad (4.5)$$

For use cases in which only the end goal is important or the feedback for sub-goals is not known, it is very common to use a single reward at the end of an episode that is propagated back to all agents. This is called sparse reward design. As we aim to achieve the minimum makespan as the global objective in our scheduling problem, it is obvious to also evaluate a sparse global reward design in contrast to the dense local and global reward design. The hypothesis is that it could be challenging for the agents to learn because no rewards are received for sub-goals. Without being explicitly rewarded for these sub-goals, it could take longer for the agents to learn a functioning policy as the sparse global reward is only received when all agents finished the episode correctly because the reward is calculated by the makespan (Pol et al., 2021). The advantage of the sparse global reward is that it is very intuitive from the modelling perspective as it can directly use the makespan and does not require an estimate of the lower and upper bound beforehand to set up the function that is used for the calculation. We calculated the sparse reward in a similar way to the R_{local} for valid machines, with the difference that we use the makespan that includes all processing times of machines, waiting times, and transportation times. We subtract the total makespan from the estimated highest makespan possible that is the compensation reward for transportation R_{comp}, and the maximum processing time $T_{max\_processing}$, both multiplied by the maximum number of operations $N_{operations}$, as formalized by the following equation:

$$R_{globalsparse} = N_{operations} \times (R_{comp} + 0.1 \times T_{max\_processing}) + (-0.1 \times c). \quad (4.6)$$

4.3   Considered Flexible Manufacturing System

As we need a representative manufacturing system to develop and evaluate the concept, we use an example FMS, where the concept can be applied and deployed to the existing production control system and MES afterwards. In the following, we introduce the FMS and a proper training environment to safely train the scheduling system.

Real Flexible Manufacturing System

The FMS in question is a modular system with six manufacturing modules that can include robots or handling systems for generalization of named machines that can be docked around the hexagon-circuit via an interface that supplies the module with load and control voltage, compressed air, and network access. Figure 4.6 shows the structure of the plant. Whenever a machine is docked or undocked, the available skills of that machine are registered at the MES. The semantic description of machine skills and the skill-matching to jobs is further described in Perzylo et al. (2019) and in section 7.2.1. Each machine can have one or more skills and can also have overlapping skills with other machines. There is only one instance per machine and either the machine is docked and "available" or undocked and "unavailable" from the MES perspective. As the number of machines $N_{machines}$ and number of

Figure 4.6 Structure of example FMS

machine interfaces are both six, the number of topologies with all machines available $T_{available}$ can be calculated by equation 4.7:

$$T_{available} = N_{machines}! \qquad (4.7)$$

where there are, in total, $6! = 720$ ways the six machines can be arranged to the six interfaces around the production circle. For training purposes, we virtually create additional machine topologies where one machine is unavailable at a time in each variant, as the scheduling system must be able to react to machines that are suddenly unavailable. When we consider that each machine can drop out, the number of dropped modules $N_{dropped}$ is six. As each machine can drop out for every machine topology, we calculate the number of topologies with dropped out machine $T_{dropped}$ using the following equation:

$$T_{dropped} = N_{dropped} \times T_{available} \qquad (4.8)$$

that is 6×720 and results in 4,320 topologies. When we add $T_{dropped}$ to the topologies with all machines that are available $T_{available}$, the total machine topologies T_{total} considered for our training purposes is calculated by equation 4.9:

$$T_{total} = T_{dropped} + T_{available} \qquad (4.9)$$

which results in 5,040 total topologies. Conveyors are used for transportation and the hexagon topology allows the products to recirculate and to return to a machine at a later point in time, e.g., in the case that it is occupied.

Figure 4.7 gives an overview of the components that can be used to manufacture a product with our considered FMS, the processes of the machines, and the example products. One product always consists of six components, namely five elements that can be of five different types (E1, ..., E5) to be placed on the baseplate (B1). The elements can be arranged in any order and rotated by 180° (except for E5 that is symmetric), leading to $9^5 = 59,049$ possibilities to configure the product (Pol, 2020). The baseplate is placed on a workpiece carrier (W1) that has handles to be grasped for transportation. The workpiece carrier is placed on a small cart (C1), which is the transport medium on the conveyor and is assigned to the workpiece carrier until the manufacturing of the product is finished. P1 is an auxiliary process that can be executed by every machine, because the workpiece carrier always needs to be loaded into the machine and unloaded again afterwards. In the middle of figure 4.7, the process of loading the workpiece carrier together with the product into the machine is shown, while the cart stays on the conveyor section until the process is finished and the workpiece carrier gets unloaded again. P2 is the main process to assemble an element to the baseplate. The processing time it takes a machine to execute a certain operation includes the time that is needed for P1 (load), P2 (actual process), and P1 (unload).

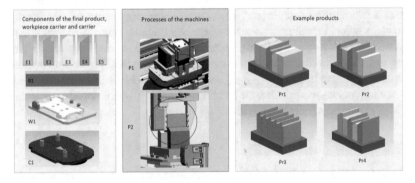

Figure 4.7 Components used to manufacture the product (left), processes of the machines (middle), and example products consisting of five elements from up to five different types that can be arbitrarily arranged and rotated (right)

As the FMS is non-productive and just used for research and demonstration purposes, there is a recycle operation for recycling the product again (P3) that is not explicitly depicted, because it is the reverse process of P2. Four example product variants (Pr1, ..., Pr4) out of 59,049 products that can be manufactured within the FMS are displayed on the right-hand side.

The operation of a job is defined by one of the five elements, its rotation, and the position to assemble or to remove the element. The skills of a machine determine if a certain operation can be executed. The machines have different materials in their material buffer and, depending on their available tools and materials, their skills are determined. To assemble element E1 and E2 with rotation 0 involves two different skills of a machine. However, to assemble element E1 with rotation 0 on position 1 would describe the skill in more detail, but we define that elements can be only assembled from left to right. Therefore we consider that every machine with assembling skills can assemble elements to any position when we start from left to right, because with this order no gaps between elements can occur and we can neglect the position of the element in the skill definition.

To highlight the complexity, we summarize: one single product can be one out of 59,049 variants to be processed in one out of 5,040 machine topologies, leading to almost 300 million setups that a product can be manufactured in. These 300 million setups are only possible boundary conditions in which a schedule needs to be found. Furthermore, we aim to find good schedules for n products at a time including machine dispatching and transportation.

The use of job specification condenses the 300 million possible setups to the number of possible job specifications in combination with the machine topologies. The job specification thereby includes the information of the 59,049 product variants in an encoded way that the agents need to learn to interpret during the training. As all machines can execute P1 and P2 and some of them can execute P3, the complexity is mainly determined by the material availability and their location docked to the FMS.

To achieve well-generalized agents, the training effort is high when performing the training in the real FMS as training is limited by the speed of the production. It would further impair the productivity when untrained agents explore situations during production to learn from successes and failures. We therefore see the need to develop a digital representation of the FMS in which agents can explore any situation without affecting the production. Furthermore, we can create an advantageous mix of situations to develop a scheduling system that knows how to interpret new situations.

Digital Representation of the Flexible Manufacturing System

The schematic of the real FMS is shown in figure 4.8A and the mapping to the virtual depiction in B is described in the following section. The conveyor sections are marked as dashed lines and the switches are controlled for transport decisions. For the example FMS, we assume all transportation times to be equal as the distances of the conveyor sections are equally long. The carts are stopped on certain positions to read their Radio-Frequency Identification (RFID) transponders and to execute the next step of the automation process, e.g., to change the switch position of the next junction for the current product. For machine interface four that is marked by a filled dot at the top of figure 4.8, there are two stoppers that are marked by small empty circles labelled by elevens with the RFID reader, one in the queue of the machine and one on the conveyor in front of the machine interface. Both locations can be condensed to one decision-making point (which is place 11 in B), as on both locations it needs to be decided whether the product should stay on the conveyor or whether it should be assigned to machine four at machine interface five. Based on the decision, the switch in front of machine interface five is controlled by the control system. Figure 4.8 B demonstrates the virtual depiction of this FMS with the six different manufacturing modules and six interface positions 1–6 to which they can be connected.

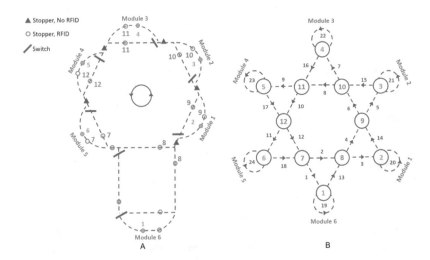

Figure 4.8 FMS (A) and Petri net representation (B) with places 1–6 as decision-making points within the modules and places 7–12 on the conveyor belt. Transitions 1–18 move products from one place to the next and 19–24 are self-transitions to stay in a module

Applied MARL Concept

We apply the presented concept to the introduced example FMS to conduct the evaluation in the following chapters based on this. Therefore, the state vector and the action space is specified, whereas the reward design is presented in the parameter study in section 5.2. One component of the state vector is the locations of all agents in the FMS. We use a 12 x 3 x 3 matrix to replicate the locations of the agents, and to provide information regarding their queue position and their ID in the first, second, and third axis respectively. The size is dependent on the queue length of each place, and on the maximum number of agents in the FMS, which we both limited to three to start with a complexity that is challenging enough to test the concept. Lastly, we include the machine topology, which defines the order of all available and unavailable machines around the manufacturing plant as a 7 x 6 matrix with six columns representing the machine interface places where a machine can be connected to the FMS. We concatenate the one-hot encoded components and flatten the resulting matrix to shape the naive state input vector that we can feed to the agent at each decision-making point. Our final vector has a length of 166 bits with the determined number of three agents and unused agents padded by zeros. For the advanced state design, we add the encoded job specifications. We therefore defined the range of 1 to 9, as the processing times of the machines in our FMS vary from 10 seconds to 90 seconds. We assume that a maximum deviation of 5 seconds is still accurate enough when we round the processing time to the nearest tens and divide it by 10 to transfer it into an integer value in the defined range from 1 to 9. The value zero is used for a machine that cannot process the operation. The processing time is also one-hot encoded, as it is added to the other one-hot encoded components, requiring 10 input nodes for each value of T_{o_i,m_m}. Although the advanced job specification encoding is designed to keep the agent's state as small as possible, thereby avoiding redundant information, it still requires substantial space. Considering three agents in the advanced job specification encoding with a lookahead of two, six machines, and ten time units for each operation requires 360 additional input nodes for the neural network. A lookahead of three would require an additional 180 input nodes. Adding the 360 nodes to the already existing 166 input nodes, the state input results in a size of 526 bits. Considering the state input to select an action from the defined action set, the agent must be able to evaluate their selected action that is done using the reward design of the following section.

For the action space variant of A_{all} actions, the agent has to select one out of 24 actions because this is the total number of transitions in the Petri net of figure 4.8. One drawback is that at each place in our considered FMS, there are two valid actions and 22 invalid actions that need to be understood and learned by the agents.

The action design with action space A_{small} consists of two actions, as there are a maximum of two transitions for each place pointing in the forward direction.

4.4   Evaluation of the Technical Functionalities

We chose the HoQ as an overall methodology to approach the objective of this work and performed step six in this section to map the requirements of the market view with the technical functionalities of the concept. To evaluate step six, we compare the requirements of section 2.2.3 with the functionalities that were developed in the proposed concept in table 4.3. We thereby also evaluate the dependencies and relations of the requirements and functionalities as step seven and eight of the HoQ. The aim is to double-check whether all requirements were considered in the concept before starting with the experiments to avoid unnecessary iterations of the concept. The experiments should then be performed with the goal of finding the best training approach that builds on this concept and to evaluate how well the concept meets the requirements. The highlighted requirements 11–14 represent the determined research gap and are therefore the most important requirements to be considered. One of the main differences to existing approaches is the assignment of one agent to each product instead of assigning them to the machines or having a central decision-maker. With the agent being assigned to the product, transportation decisions including transportation times (7) can be considered in addition to the machine dispatching. The desire for less engineering effort (15) is addressed by the general concept of training self-learning RL agents in an environment that can be set up with less effort. The training takes place in a safe environment using a Petri net as the digital representation of the FMS that is defined by the incidence matrix considering the plant topology and the complex material flow (11). The state that is given to the RL agent is designed to consider the overlapping skill-sets of machines (16) and machines with multiple skills (19) by including the job specification. With the generic encoding of the job specifications that describe how the product can be manufactured, including the available machines for each operation, the agents should be able to deal with a large number of product variants (12) and unseen product variants (13). Our approach is to filter all available machines able to process the operations of a job for the job specification included in the state description for the agent. This entails that the RL agents do not directly learn the skills of the machines, but rather use the context information of the state representation (14). Requirements (12) and (13) are further evaluated by developing a proper training strategy and the supply of training samples in chapter 5. The action design in combination with the Petri net approach is set up in such a way that the agents navigate the products to

Table 4.3 Technical functionalities of the concept to address the customer requirements

ID	Requirement	Functionality
12	Large amount of product variants	Generic description of job specification
13	New products with less effort	
16	Overlapping skill sets of machines	Considering available machines in job specification
14	New machine skills with less effort	Considering available machines + random sampling
17	Multi-objective optimization	Decentral MARL + local reward design
18	Global optimization goal	Global reward design
6	Reaction to unknown situations	Generalization
5	Reaction to unforeseen situations	Situation-based decision-making
11	Complex material flow	Petri net simulation
15	Less engineering effort	Self-learning MARL + Petri net

the machine interfaces where a machine can be docked (14). With this concept, the hypothesis is that new machines (depending on the number of different machines used for the training) or machines that receive a new skill by incorporating a new tool or material can be handled by the agents without the need of re-engineering if enough randomized training samples are used in the training phase. With the decentralized MARL approach, the problem is distributed to independent decision-makers that should be able to react situationally whenever necessary (5) and through the use of NN the hypothesis is that they will learn to react to unknown situations by interpreting the information and mapping it to the most obvious action (6). With this design decision, different optimization objectives can be assigned to the agents (17) that must manufacture the product following their objective while considering other agents to fulfill the global objective in parallel (18), which is modelled by the local and global reward design respectively. Conducting step six and eight of the HoQ, the technical functionalities of the proposed concept could successfully be mapped to the customer requirements. For step seven, no conflicts are found that must be considered with special care in the following. This indicates that the concept has the necessary prerequisites to be used to find suitable training set-ups and training strategies that are investigated in chapter 5.

4.5   Summary

The overall concept was presented to train RL agents that control products including transport decisions and assignments to machines. We formalized the problem as an MDP including possible action, state, and reward designs and successfully mapped the derived requirements to the technical functionalities of the solution, thereby answering RQ1.

For the training, we use a Petri net as a virtual representation of the FMS with the advantage of fast calculation of the next state and being able to efficiently create various training samples of different situations. Agents are called at each decision-making point to choose an action to move the controlled product to the next decision-making point. Thereby, the action space A_{all} considers all possible transitions of the Petri net, whereas A_{small} considers the maximum number of possible transitions of all decision-making points. The state includes necessary information to describe the situation, such as the location of all agents, their progress and the current machine topology. For the interpretation of the required machines of all agents in the current step and for the next n time steps, the job specifications are encoded using a defined lookahead of the future. We developed two possible reward designs that are a dense global reward and a sparse global reward design. The dense global reward design adjusts the dense local reward by a global reward factor, considering the time of all agents used for transportation, waiting time in the queue and the processing time within the machines. The sparse global reward design propagates back a single reward signal at the end of each episode to all agents considering the total makespan. Both designs have the goal to achieve a cooperative agent behavior to fulfill the global objective of a minimized makespan. We successfully validated that the functionalities of the concept meet the requirements and answered RQ1. The example FMS presented is used for the evaluation o fthis in the following chapters.

Multi-Agent Approach for Reactive Scheduling in Flexible Manufacturing

5

Figure 5.1 gives an overview of the evaluation process that is inspired by the design science research cycle of (Hevner, 2007) and the evaluation proposal of customer requirements of (Huang et al., 2011). The customer requirements were collected and weighted in the first step in chapter 2. Through the conceptual design schemes, the customer requirements were considered and embodied in chapter 4. The goal of this chapter is to evaluate the conceptual design schemes by experimental evaluations in an iterative loop and to adjust the designs where needed. This process is framed by the box in figure 5.1 and is the first part of answering RQ2, where the second part is the quantitative evaluation of the solution in chapter 6. The experiments and results described in this chapter were already published by Baer et al. (2019), Baer et al. (2020b), Pol (2020) and Mohanty (2020).

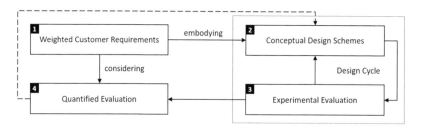

Figure 5.1 Design cycle

Supplementary Information The online version contains supplementary material available at https://doi.org/10.1007/978-3-658-39179-9_5.

S. Bär, *Generic Multi-Agent Reinforcement Learning Approach for Flexible Job-Shop Scheduling*, https://doi.org/10.1007/978-3-658-39179-9_5

A crucial component of the design scheme is the definition of proper training set-up with suitable hyper-parameters that are determined and evaluated in section 5.1. We specify the reward design empirically in section 5.2 and determine training strategies to learn meaningful policies in section 5.3. A further component for the answer to the RQ2 is to cope with a large number of product variants and new situations. We therefore develop strategies to generate and present training situations to the agents in section 5.4.

5.1   Training Set-up

It is usual to organize the training into epochs and episodes with the purpose of being able to change the environment settings after a defined period. We can perform smaller changes of the setting after every episode and certain bigger changes after n epochs. We defined 1,024 episodes within one epoch because through an experiment with different epoch sizes we found that this gives the agents enough time to explore the new setting. Following the concept of replay memory introduced in section 3.2.2, experiences of $\langle S_t, A_t, R_t, S_{t+1} \rangle$ are stored during the episodes, from which experiences are randomly sampled to update the network of the agents after a defined period of time.

As it is commonly used with off-policy learning and is easy to implement (Long-Ji Lin, 1993), we defined that the policy be updated after every epoch, which gives the agents time to explore the setting with a constant policy within the epoch. The collection of past experiences can be seen as an action model that maps states to actions together with a probability distribution of received rewards for actions in certain states (Long-Ji Lin, 1993). With the random sampling of past experiences, the agents should remember previous situations and at the same time it breaks correlations between situations. The number of products during an episode can change, as jobs are removed from the order stack if an agent has chosen an invalid action or the product was finished. When all agents are either finished or removed, the episode ends (Baer et al., 2020b).

The initial state of an episode is defined by one selected machine topology, with n job specifications for n jobs to be processed and the prioritized order to release the products. The first step was to find suitable hyper-parameters to achieve a stable training. For these experiments, we started with a limited complexity that considers a fixed order stack of three identical jobs following their local optimization goal (makespan) in one fixed machine topology. The naive state representation was used for the following experiments. To give every agent the chance to enter a machine first during training, the order to release the jobs was the only component of the initial state that we changed. Therefore, there were six initial states when shuffling the starting queue position of the three agents. In early experiments we did not shuffle the release order, which resulted in one agent always outperforming the other agents, as this agent always started first and consequently went into the machines first and let the others wait, as the agents did not yet consider a global objective. We considered three jobs represented by the job specification in table 4.1, with four operations and two machine options per operation. The jobs were released one after another to the FMS with the machines' topology $(1, 2, 3, 4, 5, 6)$, where all machines were available in the order as depicted in figure 4.2. Three agents were therefore acting in the virtual environment in parallel with the same goal of finishing their jobs as soon as possible. We conducted one experiment that changed the prioritization of job releases after every episode and a second experiment that changed it after every epoch. Figure A1 in appendix A of the Electronic Supplementary Material (ESM) compares both experiments and shows that shuffling the job releases after every episode results in agents that perform equally well, as they all get the chance to enter the machines first within one policy update that is performed after every epoch. When shuffling the job releases in every epoch, one agent slightly outperformed the other agents and the learned policies diverged slightly. We therefore changed the prioritization of job releases after every episode by default for all following experiments.

The results described in the following section were published in (Baer et al., 2020b). The training started with all agents using an epsilon-greedy policy, where epsilon defines the share of choosing random actions for exploration instead of following their learned policy. Epsilon was set to 0.99 at the start of the training, so that the agents explored during 99% of their time, which enabled them to observe various scenarios and to quickly get to know the environment. As this approach also involves the aspect that agents choose actions that might not make sense in certain situations or are even invalid, the epsilon-decay factor was used that decreases the epsilon over the training. Epsilon decay was initially set to 0.992, so that after 300 epochs the agents were following their learned policy 95% of the time and randomly exploring 5% of the time. Epsilon was decreased by the epsilon-decay factor after every policy update. The epsilon-decay factor therefore needs to be adjusted depending on the number of epochs per training (Baer et al., 2020b). Further hyper-parameters include the discount factor γ and the learning rate α introduced in section 3.1, as well as the batch-size of experiences used from the replay memory for each policy update. We empirically compared different learning rates and discount factors for the default values of $\epsilon = 0.99$, epsilon-decay $= 0.992$, and batch size $= 512$. We conducted all our experiments for this work with three different seeds to prove that it can achieve the same results for a different network weight initialization, which is one component that is dependent on the random factor. However, only the result graph for one seed is shown here for illustrative purposes.

Figure 5.2 (a) shows the effect of different learning rate and discount factor combinations. For better visualization, only one agent is displayed per hyper-parameter combination. The progress of the cumulative reward is opposed over the epochs during the training. A high gamma means that the previous state is only slightly discounted, which is beneficial when solving MrFJSP because every decision can have a high impact on the total makespan. Nevertheless, if γ and α are too high, training gets unstable, which is characterized by the jagged peaks in figure 5.2 (a). Training becomes stable with a learning rate $\alpha = 0.0001$ and discount factor $\gamma = 0.95$. For both parameters $\alpha = 0.0001$ and $\gamma = 0.95$ and replay memory size of 3 × 512, different batch sizes are compared in figure 5.2 (b). The results of batch size = 512 and batch size = 1,024 are very similar and perform better than batch size = 64.

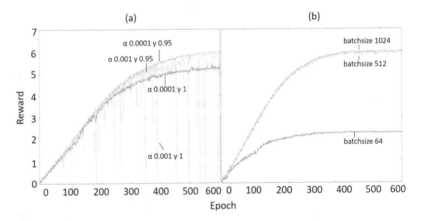

Figure 5.2 (a) shows the comparison of different learning rates and discount factors. The combination of $\alpha = 0.0001$ and $\gamma = 0.95$ has the most stable learning progress. (b) shows the comparison of different batch sizes with $\alpha = 0.0001$ and $\gamma = 0.95$. Batch size = 512 and batch size = 1,024 performed equally well (Mohanty, 2020)

Figure 5.3 shows the stable training progress of all three agents using the best parameters of the presented experiments. After approximately 500 epochs, the agents converged to a policy that no longer improved. Besides evaluating the accumulated reward during the training progress, the agents were tested using the job specification that was presented during training and were able to find the optimal schedule for the six different initial states. For the hyper-parameter study, it was reasonable to conduct the experiments with very limited complexity. The second step was to find a suitable strategy to present various situations to the agents during training, which most likely would require a longer training.

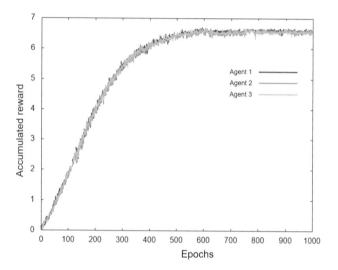

Figure 5.3 Training progress of three agents trained with $\epsilon = 0.99$, $epsilon\_decay = 0.992$, $\alpha = 0.0001$, $\gamma = 0.95$, $batchsize = 512$, replay memory size $= 3 \times 512$

We investigated different epsilon-decay factors that are compared in figure 5.4 for a 2,000 epochs training duration. This factor should not be too small (such as 0.995 and smaller) so that the agents have only a small amount of time during the training to explore, and not be too high (such as 0.999 and higher) so that the agents do not even follow their own policy towards the end of the training. With an epsilon-decay of 0.998, the agents exploit using their own policy after 80% of the training with a probability of 95%, which means that there is 20% of the training left to evaluate and fine-tune the learned policy. For the reasons outlined, we defined $epsilon\_decay = 0.998$ as a default parameter for subsequent experiments.

The experiments shown were performed using the action space of 24, as this was the most straightforward approach to use all transitions of the Petri net as the action space in every situation. Unavailable actions were learned quickly by the agents in the early training phase (Baer et al., 2020b). We also trained the agents with an action space of two, which did not yield better results. The full action space of 24

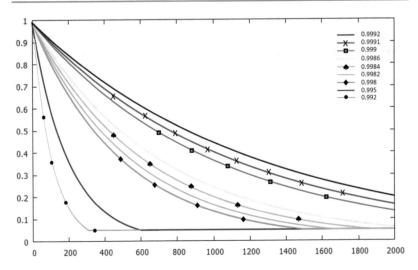

Figure 5.4 Progress of different epsilon decays over 2,000 epochs

actions was also selected so that the meaning of each action stays the same and is not situation-based, as was the case when there were only two actions to choose from. This might be beneficial for the learned network structure because some actions are not interesting for designated state components, as is the case for some locations in the FMS.

5.2   Specification of the Reward Design

The reward concepts that were presented in chapter 4 are generic and need to be adjusted depending on the FMS they are applied to. We therefore conducted parameter-studies and performed experiments to fine-tune the local and global reward designs for the considered example FMS. We further describe how the reward designs are applied during training. This is a crucial component to answer the RQ2, as the reward concept and strategy are the basis for cooperation that is again crucial to achieve a common optimization goal.

Specification of the Local Reward Design

We conducted reward parameter studies to empirically define the explicit values for the dense local reward function of equation (4.2) as follows (Pol et al., 2021):

$$R(s, a, s') = \begin{cases} -0.1 & \text{valid transportation steps} \\ 1 + 1 + (-0.1 \times (T + W)) & \text{valid machines} \\ -1 & \text{invalid steps} \\ -0.8 & \text{invalid machines} \end{cases} \quad (5.1)$$

In the parameter studies, we used various values for the variables R_{comp}, $T_{max\_processing}$, $R_{transport}$, $R_{inavlidsteps}$ and $R_{invalidmachines}$. The best results were achieved with $+1$ for the compensation reward R_{comp} for the maximum number of five transportation steps to reach a machine. Starting with $R_{comp} = +0.5$ in the experiments, the agent was not motivated to go for the machine that is far away, so we iteratively increased R_{comp}. We used $+1$ for $T_{max\_processing}$ because the maximum processing steps are 10. The processing time T that is inverted and scaled in the range of $[0, 1]$ was added to motivate the agent to choose the faster machine. An example of a local reward of an agent that selects a machine that needs three time units to process the current operation would be 1.7 using equation (5.1). Valid steps refer to the transportation steps on the transportation system or to the waiting time in the queue of a machine and therefore $R_{transport}$ is defined as -0.1 to reward it with the same importance as the processing time in a machine. An invalid machine is not able to process the operation and an invalid step is a transition that is not allowed in that specific place (Pol et al., 2021). If the selected action led to an invalid transition, $R_{inavlidsteps}$ of -1 was received for this action and the agent was removed from the simulation. The same applied for choosing an invalid machine which was penalized with $R_{invalidmachines}$ of -0.8. The differentiation between invalid step and invalid machines was made because while the transition that is chosen by the agent when entering an invalid machine is valid, the machine itself cannot process the needed operation. To choose an invalid transition that is rewarded by $R_{inavlidsteps}$ of -1 should thus be avoided.

Specification of the Global Reward Design

The global reward design was empirically defined by experiments with different constants. By analyzing the resulting total makespan of a number of experiments for various calculated upper and lower bounds, we found that suitable constants used in equation (4.3) were $w_1 = 1$, $w_2 = w_4 = 0$, $w_3 = 1.1$. By means of the constants, the upper bound estimations were adjusted by an increase of 10% to consider timings of agents using the same machines followed by waiting times that need to be considered. The lower bound did not need to be adjusted and resulted from the maximum of lower bounds of all considered job specifications. For the global reward factor, we found that the constants $f_1 = 3$, $f_2 = 0.2$, $f_3 = 1$, and

$f_4 = 0.8$ worked best to successfully train agents that aim for fast processing of their own jobs in addition to a minimum total makespan. Figure 5.5 shows an exemplary function for a lower bound of 30 and an upper bound of 60, where the label Mean indicates the mean value between the two bounds. For a total makespan lower than the Mean, the global reward factor was higher than +1, and for a total makespan below the Mean the global reward factor was lower than +1, but at a minimum of 0.8.

Formally, the function for the global reward factor $R_{factor}(c, f_1 = 3, f_2 = 0.2, f_3 = 1,$ and $f_4 = 0.8)$ is defined by the following equation (Pol et al., 2021):

$$R_{fac} = \begin{cases} 3^{\frac{m-c}{m-b_{LG}}} & \text{if } c \leq m \\ -\frac{0.2}{b_{UG}-m}c + \frac{0.2m}{b_{UG}-m} + 1 & \text{if } m < c \leq b_{UG} \\ 0.8 & \text{if } c > b_{UG} \end{cases} \qquad (5.2)$$

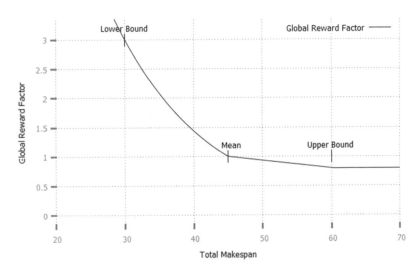

Figure 5.5 Function used to map the total makespan to a global reward factor given the global upper and lower bound (Pol et al., 2021)

As defined in the MMDP in section 3.3, all agents received the same reward. This is true for the global sparse reward that can be only calculated at the end of the episode and is given to all agents. However, the logic of feeding back the dense global reward of equation (4.5) works slightly differently. The global reward factor is calculated at the end of the episode and the local dense rewards that are stored in the replay memory during the episode get adjusted by the global reward factor at the end of each episode. Therefore, depending on the local reward, the reward of each agent can differ. The adjustment of the dense local reward by the global reward factor is only performed for state-action pairs that directly affect the global reward. We iteratively found that boosting negative local rewards leads to defensive agent behavior and thus defined using equation (4.5) that negative local rewards are not adjusted. As determined by equation (5.1), this includes local rewards for transportation steps, invalid steps, and invalid machines. For example, steps on the conveyor are just a means of transport and should not be rewarded higher or lower based on the overall result. This was a finding after analyzing the first experiments and the behavior of the agents. When adjusting the rewards of the transportation steps, agents sometimes tend to prefer to stay on the conveyor rather than choosing a machine. When excluding these state-action-pairs from the adjustments, this defensive behavior stopped. The local reward for entering valid machines can also become negative if the waiting time outweighs the compensation reward $R_{comp} + 0.1 \times T_{max\_processing}$. We observed that the agents quickly learned to avoid negative rewards and that these thus do not need to be additionally boosted. Therefore, only positive local rewards were adjusted by the global reward factor, as they have a relevant effect on the global reward and the cooperative agent behavior.

The calculation of the sparse global reward defined by equation (4.6) can be explicitly demonstrated with $R_{comp} = 1$, $T_{max\_processing} = 10$ and the number of operations $N_{operations} = 4$ by the following equation:

$$R_{globalsparse} = 4 \times (1 + 0.1 \times 10) + (-0.1 \times c)$$
$$R_{globalsparse} = 8 + (-0.1 \times c). \tag{5.3}$$

Application strategies of Rewards

After brainstorming of how to combine the local and the global reward, we summarized five possible strategies in table 5.1. We already published and explained the methods in Pol (2020) and Pol et al. (2021). The first application strategy directly combines the local and global reward functions. However, agents explore and make mistakes in the early training phase, which makes it difficult to obtain a makespan to calculcate the global reward. It is therefore useful to split the training into two phase, to give the agents the chance to learn basic behaviors in the training phase and to fine-tune the behavior in a re-training. With strategy two, the re-training is solely conducted using a global reward function, which can lead to a complete change

Table 5.1 Reward Strategies (Pol, 2020)

Strategy	Training	Re-Training
1	Local Reward + Global Reward	–
2	Local Reward	Global Reward
3	Local Reward	Over-fitting with current Jobs
4	Local Reward	Local Reward + Global Reward
5	Global Reward	–

of the learned policy, if the global reward functions differs from the initially used local reward function, which is inefficient and therefore not our favoured strategy. We already tested this strategy and published the results in (Mohanty, 2020). With strategy three the re-training is performed using the actual jobs right before the production starts. This might lead to a good result for the current jobs, as the policy gets over-fitted for the jobs, but the strategy does not follow the goal to generalize to a large number of product variants or to unforeseen situations during run-time. For these reasons, the strategy is not selected for our concept. Strategy four combines the local and the global reward function in the re-training, which mitigates the problems of the strategies mentioned before. By the combination of local and global rewards, the agents get fine-tuned by the adjustment of the rewards they know from the training phase. For this reason, we decide to use training strategy four for our concept. Therefore, we utilize the local reward function of equation (5.1) for the training phase. In the re-training phase, the local rewards stored in the replay memory are adjusted retroactively at the end of an episode when the makespan is known using equation (5.2). In the case that the episode can not be finished successfully, the global reward factor is set to the value of one by default, which would leave the local reward unaffected. To cope with this issue, we adjusted the implementation so that the agents could almost make no mistakes during re-training. When an agent does an exploratory action, only valid actions are filtered using the job specification and the location of the agent with respect to the valid transitions in the Petri net. From that valid transition, the exploratory action is chosen using the random function. When an agent performed an exploitative action, it could still happen that an invalid action was chosen, as we were using the action space 24. However, the agent should have learned to distinguish between valid and invalid actions in the early phase of the training, so that this happened rarely during the re-training (Pol, 2020). With this adjustment, we ensured that all agents finished most of the episodes during re-training and the global reward factor could be calculated correctly (Pol et al., 2021).

Strategy five uses the global reward function only without a re-training, which is the strategy that we use for the sparse global reward approach. A single reward is received by all agents at the end of an episode. It can be difficult for the agents to derive their contribution from a single reward signal at the end of an episode. To deal with the known credit-assignment problem, eligibility traces Sutton and Barto (2010) are used and replace the experience replay memory. Eligibility traces take into account the past states when propagating back the learning of $t + 1$, unlike Q-Learning where the action-value function is only updated for the state for the time steps t and $t + 1$. This approach is needed to also adjust the value of the first state-action pair of an episode, when solely a single reward signal is received for the last state-action pair of an episode. In the eligibility trace $et(s)$ the state s for each time step t is maintained. When a network update is performed, the current state and all previously visited states are updated depending on the size of the eligibility trace. The eligibility trace faces for every time step a state is not visited. As the order of state visits is crucial for the network updates, experience replay can not be used anymore, as it randomly samples tuples with the goal to break the correlations, which is counterproductive. With eligibility traces, the stored trajectories are sequentially processed for the Q-value updates.

As the sparse reward can only be calculated, if the episode was finished correctly, we also need to apply a filter for the sparse global reward approach, similar to the filtering of valid actions for exploratory actions using the dense global reward approach. Especially in the cases, where agents are making mistakes, the agents would not get the chance to improve, as the reward can not be calculated properly. Therefore we ensure that the agents can only select valid actions by applying *Q-value masking* similar to Kool et al. (2019) and Bello et al. (2017). Invalid actions of the action space are masked out. During exploration, valid actions are selected randomly and during exploitation the agent values all actions and we select the highest Q-value from the valid actions. If an invalid action has the highest Q-value it is disregarded. With this concept, we ensure that every episode can be successfully finished by all agents fulfilling their jobs correctly.

5.3   Evaluation of Suitable Training Strategies

The training strategy consists of a suitable MARL approach, a suitable learning method, and an approach to present situations to the agents during training. After selecting a learning method from the learning methods presented in section 3.3, the

evaluation of these aimed to achieve a stable training. The overall goal of defining the training strategy is to prove if the concept that was defined so far enables
agents to generalize their policy to unseen experiences. We thereby considered the
requirements of a large number of product variants (12), new product variants (13),
and overlapping (16) and new skill-sets of machines. We evaluated the training
approaches to present the situations to the agents using hundreds of job specifications, including unseen ones.

5.3.1 Evaluation of MARL Algorithms

Different RL approaches were considered and evaluated. Deep Deterministic Policy
Gradients (DDPG) was introduced by Lillicrap et al. (2015) and has an actor-critic
structure that utilizes one actor network and one critic network. The actor network
selects the action using the policy gradient algorithm and one critic network evaluates
the action in a similar way to the DQN approach by estimating the Q-Value for
this action. Lowe et al. (2017) proposed the multi-agent policy gradient algorithm
Multi Agent Deep Deterministic Policy Gradients (MADDPG) that performs well
in mixed cooperative-competitive environments. The multi-agent version expands
the critic network by the observations and actions of all agents, which results in one
central critic network estimating the Q-value of the joint actions that is fed back
to the individual actor networks. This concept enables a situation-based evaluation
of actions, not only by receiving the reward, but also by understanding the other
agents' observations and actions. Policy gradient algorithms directly map states to
actions by estimating the gradient of the expected return over the episode for that
specific action instead of estimating the probability distribution across all actions.
Therefore, they can only be used for continuous action spaces. However, Samsonov
et al. (2021) successfully applied the policy gradient algorithm Soft Actor-Critic
(SAC) for the order release and operation sequencing within an FMS. Their action
space design is flexible in terms of the number of jobs and operations per job and can
be used by RL approaches with discrete and continuous action space. The action
space is defined as $[-1, 10]$, representing the processing time that afterwards is
mapped to a specific operation. Using the continuous action space of SAC, the
agent selects an action from the continuous action space that is mapped to the
operation with the processing time that is the closest to this value. Kuhnle et al.
(2020) designed a similar mapping of the continuous action space to the actual
discrete actions using Trust Region Policy Optimization (TRPO) and (Roesch et al.,
2019) successfully applied Proximal Policy Optimization (PPO) to the scheduling
problem at hand, which are both policy gradient algorithms with continuous action

spaces. Inspired by the concepts of Kuhnle et al. (2020), Roesch et al. (2019), and continuous action space design of Samsonov et al. (2021), together with the promising approach of Lowe et al. (2017), we implemented the MADDPG approach and adjusted the continuous action space to our discrete action space by defining the output range and mapping the range to the corresponding discrete action afterwards. Although this approach worked well for the order release of Samsonov et al. (2021), in our case the agents did not learn how the actions are mapped to the transitions within the Petri net and therefore did not learn meaningful policies. Furthermore, the training was very unstable, the policies did not converge, and the agents seemed to be confused by the continuous action space. We conducted several parameter studies for the learning rate, discount factor, and the soft-update coefficient that determines the policy update frequency, and also tried different network sizes but could not achieve a stable training. One reason could be that in our approach, the continuous action space is mapped to concrete transitions in the Petri net, which requires a very precise action selection, because a deviation in the action can result into a different transition. For the approaches, where the actions represent processing times, the deviation has not such a huge impact as when mapping them to transitions, which result into completely different states than the agent probably had expected. Although, the MADDPG seems reasonable for some approaches, the comparison to benchmark RL approaches of Lowe et al. (2017) is lacking because it was compared to DQN but not compared to Multi Agent Deep-Q-Network (MADQN). The DQN approach proved to work well for our less complex scenarios in section 5.1 and was also successfully used by Waschneck et al. (2018) and Qu et al. (2016). Even if the MADDPG approach did not perform well for our MrFJSP, we regarded the concept of cooperation utilizing the information of the other agents to be beneficial and therefore adapted the concept to the MADQN approach. This inspired us to design the state space as shown in section 4.2.2 by sharing the necessary information via the state vectors. However, we did not share the actions of other agents yet, as we only used one network that performs the action selection and the evaluation at the same time. For the aspect of cooperation we designed our dense and sparse global reward functions in a way that should motivate the agents to collaborate to achieve their common objective. With this MADQN approach, we did not use explicit exchanges of information or communication protocols in the training or in the decentralized execution during runtime. Each agent had to sense the actions of the other agents from the state information provided, as the actions affect each other and are not explicitly known.

5.3.2   Selection of MARL Learning Methods

We assessed the methods that were introduced in section 3.3 based on our requirements and selected methods that we used for the experiments. We decided against
JAL (Claus and Boutilier, 1998) and (Bloembergen et al., 2015), as the network
must be trained from scratch for every additional agent, which makes this approach
inflexible. Furthermore, the size of the action space grows exponentially with the
number of agents. The decision-making of single agents is further dependent on
their location in the FMS and does not need to be synchronous with all agents. For
independent decision-makers, the approach of using decentralized agents or IL is
more suitable. IL are independent of the number of agents and more flexible, as
sub-policies with different optimization goals can be trained. However, IL do not
cooperate by default as they consider other agents only due to the changing environment and are only trained using their own experiences. Concurrent learning aims
to reduce the complexity of the search space and computation expenses, especially
for problems that can be decomposed into smaller tasks with heterogeneous agents
that cooperate to fulfill the common goal. In the MARL environment, the behavior
of the system is affected by agents that run simultaneously and act independently.
When running multiple agents in parallel in the MADQN setting without explicit
exchange of information, it can become difficult for the agents to learn something
due to the resulting non-stationary environment. To overcome this issue, Maxim
(2016) recommends letting all agents act in parallel while only updating the policy
of one agent using the experiences of all agents. The policies of the other agents
are fixed and get updated after a defined number of epochs. This learning method is
called parameter sharing and helps to overcome the issue of a non-stationary environment. It is one concept that could meet our requirements of achieving cooperating
agents that aim to minimize the makespan. Figure 5.6 shows independently acting
agents using the MADQN approach that follow a common optimization objective
and can have a local optimization objective and respectively reward functions when
this information is added to the state vector. The architecture shown represents the
parameter sharing learning method, as individual networks $Q1, .., Q3$ are used for
the agents.

As we considered homogeneous agents, we can leverage the centralized learning
version of Gupta et al. (2017a) and train one network that can be used as independent
sub-policies for each agent (Gupta et al., 2017a). We slightly adapted the concept
of Gupta et al. (2017a), as they suggested using the joint observation and separate
action spaces. As the action space is identical for all agents, the same network
with the same action space can be used although the observation differs slightly
for each agent as we followed the advanced state design presented in section 4.2.2.

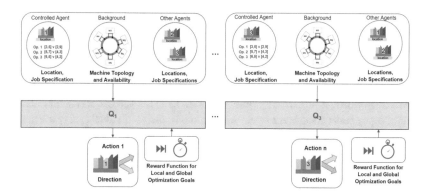

Figure 5.6 Agents act independently using the MADQN approach, while following a common optimization objective. The state vector includes information about the controlled agent, the environment and the other agents that should help the agents to anticipate the actions of other agents in order to cooperate

We eliminated the non-stationary environment, as we only utilized one network $Q1$ and assumed that the environment stays stationary between the policy updates. As all agents can profit from the experiences of other agents when storing them in a collective replay memory (Gupta et al., 2017a), we aimed to achieve the same advantages as parameter sharing.

Learning of Coordination and Coordinating Q-Learning were not applied as we considered a number of agents with which it was feasible to share the information directly by the state. When scaling the approach to a large number of controlled products at the same time, both approaches are worth considering for future work. We investigated how parameter sharing and centralized learning can be applied to achieve a reactive job-shop scheduling solution. We first aligned our approach to that of Maxim (2016) and applied parameter sharing whereby we trained agents with independent Q-networks and a common replay memory.

5.3.3   Evaluation of Parameter Sharing and Centralized Learning

For the following experiments, we used the same scenario as in section 5.1 and the hyper-parameters used for the experiment in figure 5.3. We defined the processing time as local optimization objectives for all agents. With these definitions, we created a slightly competitive scenario as all agents aimed to finish the job as quickly as possible (Baer et al., 2020b). The parameter crm determines whether a collective replay memory or an individual replay memory is used. If crm was set to one, we stored experience tuples $\langle S_t, A_t, R_t, S_{t+1} \rangle$ of all agents in a collective replay memory during training and sampled random experiences to update the Q-function after each epoch as proposed by Mnih et al. (2016). If the parameter crm was set to zero, the experiences of each agent were stored in their individual replay memory that was used for the policy update of the agent. The parameter cp demonstrates if *copy policy* is active, which is only the case with parameter sharing as a learning method. We compared the two training strategies of using parameter sharing with fixed agents where $Q_1 \neq Q_2 \neq Q_3$, and centralized learning with only one instance where $Q_1 = Q_2 = Q_3$.

We published these results in advance in Baer et al. (2020b) and Mohanty (2020). Figure 5.7 (a) shows the training progress of parameter sharing, where the policy of the learning agent is copied to the two frozen agents every 50 epochs. The darker graphs, labelled by "cp 1 crm 0", demonstrate the training progress using parameter sharing without collective replay memory. This means that the learning agent only learns from its own experiences and does not consider the experiences of the frozen agents. Although the policy of the learning agent is copied to the frozen agents, both performed worse, which is due to the fact that the observation of the learning agent does not match the observation of the frozen agents and the policy is not yet sufficiently generalized to interpolate the observation of the agents to predict meaningful actions. The lighter graphs in (a), labelled by "cp 1 crm 1" used the collective replay memory, where experiences of all three agents are stored, which results in a better performance of the frozen agents (Baer et al., 2020b). Figure 5.7 (b) shows the centralized learning method, using only one network for the decision-making of all agents, so cp is set to zero. The darker graphs, labelled by "cp 0 crm 0", demonstrate the training progress with $crm = 0$, meaning that only the experiences of one agent are used for the network update. This set-up performed poorly for the two other agents and does not make sense because when using only one network for all agents, all experiences must be considered for the network update. The lighter graphs in (b), labelled by "cp 0 crm 1", show the most stable training progress of the four experiments. Centralized learning in combination with the collective replay memory uses one network that is updated by the experiences of all three agents (Baer et al., 2020b).

We found that centralized learning is even more stable than parameter sharing when utilizing a collective replay memory. This learning method features the benefits of parameter sharing to overcome the non-stationary environment by the use of only one network that is used n times for the decision-making of n individual agent

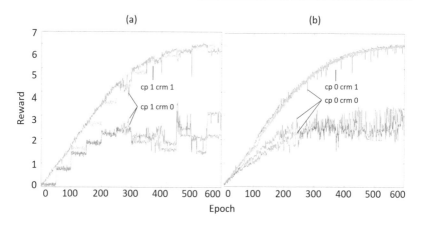

Figure 5.7 Training progress of the two learning methods parameter sharing (a) and centralized learning (b). In (a), only one agent is trained and the weights are copied to the other agents after every 50 epochs (cp = 1). In (b), only one network is trained which is used for all agents (cp = 0). The experiments were performed with (crm = 1) and without (crm = 0) collective replay memory (Mohanty, 2020)

instances. Although, the learning method is called centralized learning, the agents still act decentrally and independently while considering the other agents through their individual state inputs. We profited from the advantages of using a single Q-network as long as we considered the same local optimization objective for all agents (Baer et al., 2020b).

5.4   Training Approach to Present Situations

To leverage the presented training strategy to the full complexity of the MrFJSP, we further needed to consider a large number of product variants (12), new product variants (13), in addition to overlapping (16) and new skill-sets of machines (14), which require an approach to present training examples to the agents in a way that they learn and interpret them but simultaneously do not forget the learned behavior. The goal is to evaluate whether the agents are able to schedule the job specifications used within the training correctly and if they generalize to unseen job specifications after the training that is a very important component for the answer to RQ2. We therefore created a randomized job specification generation process and evaluated three different approaches to present these job specifications to the agents in the training. We published the experiments and results in (Baer et al., 2021) and (Pol, 2020) and summarize the three training approaches to present the job specifications to the agents in figure 5.8.

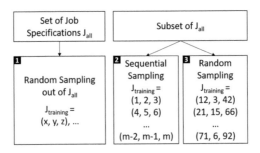

Figure 5.8 Three training approaches to present situations: random sampling from the set of all job specifications, sequential sampling from a defined subset, and random sampling from a defined subset based on Pol (2020)

For all training approaches, all agents received different job specifications, which were changed after a defined number of epochs depending on the job frequencies. We set the job frequency to one, three, and five for the first experiments. The first approach was random generation, which means that entirely new job specifications were randomly generated and assigned to the agents. The number of job specification used depended on the number of epochs during training and the job frequency. For a job frequency of one, a training length of 1,000 epochs and three agents were used, while 3,000 job specifications were used for the training and every agent would see 1,000 different job specification combinations.

Utilizing the process to generate randomized job specifications, we created a training set of 300 job specifications as our hypothesis was that this amount of job specifications is representative enough for the agents to understand the job specification and to be able to interpret it, which is necessary for understanding new ones. The second approach was sequential sampling from the training set, such as $(1, 2, 3)$, $(4, 5, 6)$, ..., $(298, 299, 300)$. Once a job specification was selected, it was removed from the training set until all job specifications from the training set were used. This concept yielded up to 100 combinations that were iterated in a loop during the training while training with three agents. For a job frequency of five and a training length of 1,000 epochs, a new combination was selected every fifth epoch thereby leading to two iterations of the 100 combinations during the training.

The third training approach is random sampling from that training set, where *n* different job specifications were randomly selected from the training set and assigned to *n* agents. With this concept, every combination of selected job specifications, i.e,. (34, 278, 16) or (97, 12, 178), is most likely different. For the job frequency of one and a training length of 1,000, the agents would see 1,000 different job specification combinations, although the same job specifications from the training set were used.

The network architecture used consisted of an input layer with the state size of 526 using the advanced state design, two hidden layers with 32 nodes, and an output layer with 24 nodes corresponding to the action-size. The hidden layers used Rectified Linear Unit (ReLu) activation functions and the output layer used a linear activation function. As we determined centralized learning as being the most suitable learning method, it was used in combination with the collective replay memory. As our focus of this chapter was set on determining a suitable training set-up, learning methods, and hyper-parameters with the overall goal to train agents that are able to generalize their experiences to unseen situations, we evaluated the results in terms of the correctness of scheduling rather than the optimality. We therefore considered the dense local reward function determined in section 5.2. The global reward function was used to fine-tune the agent behavior to achieve the near-optimal schedules that are discussed in chapter 6. We used the best hyper-parameter of the parameter study in section 5.1 $\epsilon = 0.99$, $epsilon\_decay = 0.998$, $\alpha = 0.0001$, $\gamma = 0.95$, $batchsize = 512$ and a replay memory size of 3×512.

For the evaluation of the training approaches, it is not sufficient to only analyze the training progress of the accumulated reward over the epochs, because every job specification combination can have a different maximum return that can be achieved. The repeating pattern of the different combinations shows an overall upward progress, which indicates that the agents improved their policy. However, it is not possible to interpret stability or network convergence from this data. We first evaluated the trained network using 30 job specifications that were used during training and split it into 10 scenarios, where three agents had to schedule three jobs. The agents scheduled all jobs correctly, so we obtained 100% correctness for the observed scenarios. To evaluate whether the trained network also generalized to unseen jobs and to scheduling them correctly, we created 30 unseen job specifications that were separated into ten validation scenarios. We manually designed the ten different scenarios with representative difficulties, such as the need for the same machine by different agents or consecutive operations that can be processed by the same machine. For the performance assessment, we developed a rating scheme that counts how many jobs are processed correctly without any mistake with a maximum rating of 30. A mistake would be to choose an invalid transition, an invalid machine, or to not end the job at all by eternally iterating on the conveyor circle which led to 0 points for the job in question. As we set the focus on correctness instead of optimality, we considered the correct scheduling of the 30 jobs as 100% and thus initially did not consider the makespan (Pol, 2020).

Table 5.2 compares the results of the training approaches for different job frequencies. The agents could successfully process 77% of the unseen jobs, as shown in the lower right corner of the table. The inner values (between 20 and 27) show the amount of successfully finished jobs and the outer values show the average result of the relevant columns or rows converted to floating numbers between 0 and 1. The first training approach using random generation had the worst performance. This might be due to the fact that the agents had no chance to exploit their learned policies and fine-tune them because they were constantly confronted with changing scenarios. Agents try to optimize their policy towards a moving target, because the maximum reward changes with every new job specification. The second training approach addressed this issue by sequentially sampling jobs from a generated training set. This gave the agents the chance to consolidate their experiences and fine-tune their policies on the repeating combinations of job specifications. The results were slightly better, although only 74% of the unseen validation jobs could be processed without mistakes. This might be due to the fact that using only 100 job specification combinations was not sufficient for the agents to generalize their policies to unseen jobs after the training. This problem was addressed by the third training strategy, random sampling, which had the best results. The agent correctly scheduled 87% of the unseen jobs. It seems to be advantageous to sample jobs from a training set, as the job specifications themselves are seen multiple times by the agents, which helps them to improve their strategy and stabilize. However, a new combination of

Table 5.2 Results comparison of different training strategies and job frequencies (Pol, 2020)

	Job Frequency			
Training Approach	1	2	3	
[1] Set of J_{all} + Random Sampling	21	20	22	0.7
[2] Subset of J_{all} (300 js) + Sequential Sampling	22	22	23	0.74
[3] Subset of J_{all} (300 js) + Random Sampling	27	25	26	0.87
	0.78	0.74	0.79	**0.77**

jobs was also used at every change, which allows the agent to better generalize to new combinations instead of using identical combinations each time (Pol, 2020).

We observed the tendency that the job frequency of five performs best. To prove this, we conducted further experiments with the comparison of job frequencies of 5, 10, 25 and 3, 10, 5 and in both cases we found that five performs best. Although the results in table 5.2 were promising and the agents scheduled 27 out of 30 jobs correctly, further experiments were conducted using the training approach random sampling to achieve 100% correctness for the unseen validation jobs. One approach was to double the number of epochs and the size of the training set. Another approach was to double the network size by increasing the size of the two hidden layers from 32 to 64. Further adjustments were to increase the relay memory size for storing more experiences and to increase the batch size to sample more experiences from the replay memory for each network update (Pol, 2020). The job frequency was set to 5 per default and the other hyper-parameters stayed the same for subsequent experiments. Table 5.3 summarizes the results for the experiments with the mentioned parameters after 1,000 epochs and table 5.4 after 2,000 epochs. The comparison was also published in (Pol, 2020). The highest results are highlighted and it can be seen that the agents managed to schedule all jobs correctly for several set-ups, which is a significant improvement to the previous set-up. One remarkable improvement was obtained by increasing the size of the neural network. The average percentage of correctly scheduled jobs improved by 9% for the training with 1,000 epochs and by 4% for the training with 2,000 epochs after doubling the nodes per layer. The extension of the training to 2,000 epochs led to an overall improvement of the average percentage of completed jobs of 3%, as the agents had more time to learn. The best results were achieved with a batch size of 4,096, as with higher batch sizes more experiences are sampled for the network update. Nevertheless, the higher batch size influenced the training duration, as the calculation of gradient descent is computationally more expensive using a larger batch (Pol, 2020). Using a larger training set yielded a 3% better result for the training with 1,000 epochs. In contrast, the results of the training with 2,000 epochs slightly aggravated, which is not directly visible in the figure due to rounding. However, the assumption was that it would be beneficial to use a larger training set with 600 job specifications for subsequent experiments to allow the agents to generalize, especially when the focus is set on the makespan optimization. Increasing the replay memory size per agent also yielded slightly better results. The optimal results could also be achieved with a replay memory size of 2,048 per agent and a batch size of 4,096, but the replay memory size of 4,096 was slightly better on average. The total size of the collective replay memory results therefore amounted to 12,288 when three agents are trained. We observed that the network capacity using 32 nodes per layer was too small to generalize to all

Table 5.3 Results of training for 1,000 epochs in combination with different sizes of training set, network, replay memory, and batch size (BS) (Pol, 2020)

Training for 1,000 Epochs								
Size of Training Set	300 Job Specifications				600 Job Specifications			
Size per Layer	32 Nodes		64 Nodes		32 Nodes		64 Nodes	
Replay Memory Size per Agent	2048	4096	2048	4096	2048	4096	2048	4096
BS = 1024	26	24	28	30	26	28	28	28	0.91
BS = 2048	27	26	29	29	29	27	29	29	0.94
BS = 4096	26	28	29	27	30	27	29	30	0.94
	0.88	0.87	0.96	0.96	0.94	0.91	0.96	0.97	
	0.87		0.96		0.93		0.96		
	0.91				0.94				**0.93**

Table 5.4 Results of training for 2,000 epochs in combination with different sizes of training set, network, replay memory, and batch size (BS) (Pol, 2020)

Training for 2,000 Epochs								
Size of Training Set	300 Job Specifications				600 Job Specifications			
Size per Layer	32 Nodes		64 Nodes		32 Nodes		64 Nodes	
Replay Memory Size per Agent	2048	4096	2048	4096	2048	4096	2048	4096
BS = 1024	27	26	29	30	28	28	29	27	0.93
BS = 2048	28	29	29	29	30	29	28	30	0.97
BS = 4096	29	30	30	30	27	28	30	30	0.98
	0.93	0.94	0.98	0.99	0.94	0.94	0.97	0.97	
	0.94		0.98		0.94		0.97		
	0.96				0.96				**0.96**

unknown situations, but that 64 nodes per layer are sufficient for learning 600 job specifications and generalizing the policy of the unseen scenarios. We conducted a further round of experiments with one hidden layer of 128 nodes and one hidden layer with 64 nodes to gain more capacity for considering additional scenarios in the future, such as changing machine topologies or more than three jobs. We achieved the same results as with 64 nodes for both layers and decided to continue with the bigger network for capacity reasons.

We can conclude that the best parameters are a network with one hidden layer of 128 and one hidden layer of 64 nodes, a training duration of 2,000 epochs, a training set of 600 job specifications, a batch size of 4,096 and a replay memory size per agent of 4,096. In addition to the parameter study, through the experiments we also verified that the advanced state encoding fulfills the requirements we aimed for. The agents were able to learn and remember jobs that are used within the training and schedule them correctly. Further, we examined whether the trained agents can schedule 30 unseen job specifications in 10 scenarios, which proves that the agents achieved generalized policies.

5.5 Summary

We achieved a stable training using the MADQN RL approach and centralized learning by utilizing a collective replay memory. Only one network is thereby used n times for the decision-making of n individual agent instances. The action space of 24 worked best using all transitions of the Petri net for every state (Baer et al., 2020b). The parameter study showed a stable training progress for the hyper-parameters $\epsilon = 0.99$, $epsilon\_decay = 0.998$, $\alpha = 0.0001$ and $\gamma = 0.95$. The best training approach to present situations to the agents was random sampling that randomly selects triples of jobs from a pre-defined training set of 600 job specifications. A new combination of job specifications was used every five epochs. We evaluated the experiments by creating 10 scenarios with 30 unseen job specifications and assessing their results using a simple rating scheme that counts the correctly scheduled jobs. We achieved 100% correctness with a $batchsize = 4,096$ and a replay memory size of $3 \times 4,096$, a network with 64 nodes per hidden layer, and a training length of 2,000 epochs. The experiments proved that the advanced state encoding provides the necessary information for the agents to understand and interpret the information and apply the learned policy to unseen situations. The agents generalized to unseen job specifications that can represent new product variants or new machine skills or both. We therefore fulfilled the requirements that we aimed for in the presented chapter, namely to cope with a large number of product variants (12), new product variants (13), as well as overlapping (16) and new skill sets of machines (14). As the goal of this chapter was to empirically determine a suitable training set-up with hyper-parameters, a suitable MARL approach and learning method, and a suitable training approach to present situations to the agents, the focus was set on correctness and not on optimality. While the dense local reward function was used in this chapter, the global reward functions will be used and evaluated in the following chapter to achieve near-optimal schedules with cooperating agents to reduce the makespan.

Empirical Evaluation of the Requirements 6

In the previous chapter, the design cycle to find the most suitable MARL approach and learning method was performed. This was an essential step for RQ2, which is also addressed in this chapter. We therefore evaluate the MARL approach based on the derived requirements. Figure 6.1 gives an overview of the evaluation process that is inspired by the design science research cycle of Hevner (2007) and the evaluation proposal of customer requirements of Huang et al. (2011). The fourth step framed by the rectangle is a quantified evaluation of the developed system. Through this, we conduct step nine of the HoQ, which is to evaluate the developed scheduling system based on the derived requirements. The experiment and results were already published by Baer et al. (2020b), Baer et al. (2021), Pol et al. (2021), and Pol (2020).

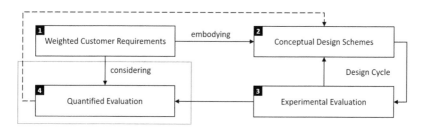

Figure 6.1 Evaluation cycle

Requirements (12), (13), (14), (15), and (16) are discussed in section 6.1, where the generalization of the solution is evaluated with the goal to be usable for various machines, overlapping skill-sets, and various and new products with less engineering effort. In section 6.2, we evaluate requirement (18) by the quantitative assessment of achievable makespans for unknown situations (6), and as a common method of the QFD (Rabl, 2009), we lastly benchmark the makespan of known and unknown situations and real-time decision-making of our solution with a heuristic search algorithm in section 6.3. Section 6.4 consolidates the evaluation results by using the Importance Satisfaction Evaluation (SIE) model for a summarized quantified evaluation.

6.1   Generalization to Various Products and Machines

In chapter 5, experiments were conducted to find out whether the state input is appropriately designed to enable the agents to generalize the job specification input and to correctly schedule jobs that were not seen during training (Baer et al., 2021). However, the validation was always performed with job specifications from the same format as those used during training. The requirement of handling a large number of products (12) and also new products (13) is likely to include job specifications of different lengths, because more or fewer operations might be needed. The requirements of considering new skill-sets of machines (14) can result in the extension of job specifications by additional alternatives. From this, the conclusion is derived that agents must be able to handle various job specification formats. The hypothesis is that the advanced state design allows the encoding of job specifications independently of their length or the number of alternatives per operation. Our goal is to evaluate this hypothesis by determining whether trained agents are intuitively able to handle job specifications of various formats without the need of a re-training on the different formats (15) (Baer et al., 2021). By using the concept of random sampling of the job specifications from a training set that was found to work best in section 5.4, we automatically included overlapping skill-sets of machines (16) in the experiments. We further concentrated on minimizing the individual makespan of the agents and neglected the aspect of minimizing the total makespan (18), as this is the focus of section 6.2. The global reward function is the key-functionality for cooperation and was not yet used in the training.

For these evaluations, we used the trained network from chapter 5 that achieved the best results. It was trained using MADQN with the centralized learning method and random sampling from a training set that consisted of 600 job specifications that always considered four operations each, with two machine alternatives. The network had two hidden layers with 64 nodes per layer and was trained for 2,000 epochs using a collective replay memory with the size per agent of 4,096 and a batch size of 4,096. The validation was conducted with 30 variably-sized job specifications that were not used in the training. Figure 6.2 shows one evaluated example result that was published in Baer et al. (2021) and Pol (2020). The agents $A1$, $A2$ and $A3$ were

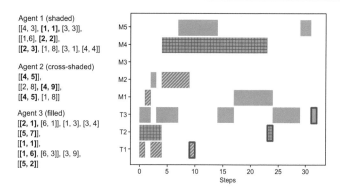

Agent 1 (shaded)
[[4, 3], **[1, 1]**, [3, 3]],
[[1,6], **[2, 2]**],
[**[2, 3]**, [1, 8], [3, 1], [4, 4]]

Agent 2 (cross-shaded)
[**[4, 5]**],
[[2, 8], **[4, 9]**],
[**[4, 5]**, [1, 8]]

Agent 3 (filled)
[**[2, 1]**, [6, 1]], [1, 3], [3, 4]
[**[5, 7]**],
[**[1, 1]**],
[**[1, 6]**, [6, 3]], [3, 9],
[**[5, 2]**]]

Figure 6.2 Correct schedule of three job specifications of various formats that were not included in the training. The chosen machines are printed in bold in the job specifications on the left. Each agent is displayed in a different color. The boxes indicate that the last operation of the job has been successfully completed

able to schedule the 30 unseen job specifications correctly. For the evaluation, the Gantt charts were plotted and analysed. The x-axis shows the makespan and the y-axis represents the available resources such as machines $M1, ..., M5$ and transport, which is displayed using three channels $T1, ..., T3$ for better visualisation of the time per agent spent for the transport. The results showed that the chosen training concept including the advanced state design enabled the agents to generalize to job specifications. We evaluated the concept using arbitrary formats with various numbers of possible machines per operation and various numbers of operations per job.

Analyzing the Gantt charts of schedules shows that the minimized individual makespans for the unseen job specifications were not always achieved. There were many cases where the agent actively skipped a slower machine to choose the faster machine. $A3$ in figure 6.2 passed $M1$ for the first operation and chose $M2$ that was two time units faster. It is important to note that the machines were docked to the FMS in sequential order $M1, M2, ..., M6$. We further observed that agents stay in a machine for consecutive operations if possible. Such behavior could be observed from $A1$ that stayed in $M2$ for the second and third operation and from $A2$ that stayed in $M4$ for all three operations. In some cases, the agents preferred the first possible machine they passed, even if it was the slowest of all possible options. The same agent chose $M5$ for the second operation, although $M6$ would have been available, took two time units less, and was only one time unit away. Although there is a slight tendency towards entering the first valid machine, the overall behavior of the agents favored the machines for their individual makespan. There are two possible justifications for the described behavior of favoring the closer, but slower machine (Pol, 2020):

- Skipping a machine that can process the current operation to go to a machine that can process the operation faster bears risks. On the way to this machine, there is the possibility that another agent enters this machine before the agent in question arrives. This would result in one extra round on the conveyor to reach the first machine again or in waiting time in the queue of the machine, which was not expected. In this scenario, the choice of the first machine would have been the better one.
- If the agent anticipates that another agent will enter the faster machine soon by interpreting the information of the job specification and location of other agents, it is likely to choose the first machine that is available rather than taking the risk. This problem results from being dependent on state information and the need to anticipate other agents' actions because there is no explicit communication of planned actions.

We can conclude that we met the requirements of handling a large number of products (12) and new products (13), as well as considering new skill-sets of machines (14) by the fact that the agents could correctly schedule arbitrary and unknown job specifications without the need for re-training. In this way, we also exceeded the requirement of a low engineering effort for adaptations of the FMS (15), as the arbitrary formats can result from new tools that are introduced to machines, enabling them with new skills, or from completely new machines that are incorporated to the FMS within the maximum number of machines the agents are trained for. We showed that agents aim to minimize their individual makespan by choosing the faster machines and by staying in machines for multiple operations if possible. Agents consider transportation costs in their decision-making, i.e., if a fast machine is far away, the agent estimates the transportation effort against the benefit of processing time. Lastly, agents avoid queues in front of machines wherever possible to reduce unnecessary waiting times (Pol, 2020). However, the analyzed schedules did not yet achieve their individual near-optimal makespan or the minimized overall makespan. The experiments showed that the agents learned to enter the first machine they pass that can process the current operation, even if it is not the fastest option, if they anticipate a risk to go for a machine that is far away (Baer et al., 2021). The goal was to encourage the agents to cooperate to achieve the global objective of minimizing the total makespan (18). In the next section, we evaluated the cooperative behavior while incorporating the adopted global reward functions.

6.2   Achieving the Global Objective

We aim to train agents that follow a global objective (18), such as minimizing the makespan. In the previous experiments, we neglected this aspect, as we focused on finding a suitable training concept and only validated the agents based on generalization. To finally achieve a global optimization, we utilized the global reward functions and evaluated the results. As we utilized de-central agents that receive their local observation, the hypothesis was that the cooperation of agents is needed to achieve global optimization. The goal is to achieve a cooperative behavior through which agents are able to recognize the needs of other agents for a certain machine and step back to give way to other agents while choosing an alternative. As the learning of the desired behavior is dependent on providing guiding feedback to the agents, we applied and compared the two global reward concepts introduced in section 5.2, in section 6.2.1, and analyzed the cooperative behavior of the agents in section 6.2.1.

6.2.1   Comparison of Dense and Sparse Global Rewards

The first approach was to utilize the dense global reward with strategy four of training and re-training introduced in section 5.2. In the first phase, agents were thus trained based on the dense local rewards with the goal to achieve individual minimum makespans. Agents were re-trained in the second phase by boosting the dense local rewards using the global reward factor with the focus on cooperation. Through the concept of boosting the rewards and not changing the value range of the reward completely, the learned policy should become fine-tuned. Training strategy five introduced in section 5.2, was to train by solely using the sparse global reward function, where a single reward signal is propagated back to all agents at the end of an episode (Pol et al., 2021). In addition to the reward functions, the advanced state design of section 4.2.2 provides the necessary information about the needs for machines that depend on the upcoming operations of other agents. The locations of all agents should help them further to interpret this information and to plan the machine usage and timings accordingly.

For the experiments, we used the training parameters that were mentioned in section 5.4. The MADQN approach using one Q-Network with central learning consists of two hidden layers with 128 and 64 nodes respectively using ReLu activation, and the output layer with 24 nodes with linear activation for the action space 24. We stored the experiences in the replay memory of 4,096 per agent and sampled 8,192 using stochastic learning. The learning rate was set to $\gamma = 0.0001$ and $\alpha = 0.95$. We trained with random sampling using the training set with 600 job specifications

Table 6.1 Results comparison between using dense local rewards for 2,000 and 4,000 epochs and using dense and sparse global rewards after 4,000 epochs (Pol et al., 2021)

| Dense Rewards | | | | Sparse Global Reward | | |
| Training for 2,000 Epochs | After Re-Training (Additional 2,000 Epochs) | Dense Global Rewards | Improvement Col. 1 to 3 | Training for 2,000 Epochs | Training for 4,000 Epochs | Improvement Col. 5 to 6 |
Dense Local Rewards	Dense Local Rewards					
37	39	41	−11%	47	42	11%
36	50	30	17%	34	34	0%
61	56	46	25%	37	38	−3%
35	31	37	−6%	31	33	−6%
41	34	36	12%	45	45	0%
32	34	34	−6%	41	32	22%
40	38	31	23%	43	39	9%
42	39	38	10%	39	38	3%
40	40	40	0%	41	32	22%
40	45	32	20%	35	33	6%
40.4	**40.6**	**36.5**	10%	**39.3**	**36.6**	7%

and a job frequency of five. We adjusted the epsilon decay to 0.9984, as we trained for 2,000 epochs and conducted a re-training of another 2,000 epochs where we again set the epsilon to 0.9 to allow exploration during the global reward training. The described experiments and the following results were published in (Pol et al., 2021).

A form of quantitative assessment is required to be able to evaluate and compare the results. For this purpose, a validation set was used for all experiments that comprised ten scenarios, in each of which three different products must be scheduled that were not used during training. The total makespan achieved for each validation scenario, as well as the average makespan of all scenarios, can therefore be compared between the experiments. Table 6.1 shows the results of the two reward concepts.

The four columns on the left side of the table represent the results of the dense reward approach. For the dense reward approach, we compared three variants. We evaluated the results using only the dense local reward after 2,000 epochs and after 4,000 epochs and re-training using the dense global reward after 4,000 epochs. We compared the three steps to find out whether the longer training or the re-training led to the policy improvement. We used the relative improvement in percentage p calculated by the following equation:

$$ p = (y_1 - y_2)/y_1 \times 100 \tag{6.1} $$

where y_1 is the base value that is expected to be smaller than the improved value y_2. Otherwise, the relative improvement is negative.

The training using the dense local reward for 4,000 epochs did not lead to improvements compared to the training for 2,000 epochs. Comparing column one and two, the makespan of four validation scenarios increased and the makespan of six scenarios decreased, while the average total makespan stayed the same. The total makespans of the re-training using the global dense reward are shown in column three. Column four shows the comparison of the results of the re-training after 4,000 epochs with the baseline training of 2,000 epochs in percentage. In six validation scenarios, the makespan improved significantly by up to 25%, one scenario did not change, and in three scenarios the total makespan became worse. The average total makespan shows an improvement of 10% compared to the baseline training, which confirms that the global dense reward had a positive effect on optimizing the schedules by minimizing the makespan.

Column five and six show the comparison of using the sparse global reward for 2,000 and 4,000 epochs. The average total makespan improved by 7% during the training process from 2,000 to 4,000 epochs. Comparing the average total makespan of the sparse global reward (36.5) and the dense global reward (36.6), similar performance was achieved. It is remarkable that the results using the sparse global reward function, which are only fed back to the agents once per episode could keep up with the dense global reward function that is well-thought out and engineered for different situations. The agents thus overcame the credit-assignment problem that is accompanied by the use of sparse rewards. The engineering of the sparse reward function required less effort than the development of the dense local and global reward functions, including finding suitable hyper-parameters. It is therefore important to understand that the dense reward functions cannot be used for different FMSs without adjustment according to the new processing and transportation times of the FMS, which requires engineering effort for adjusting the dense local reward of equation (5.1), the dense global reward factor (5.2), and the fine-tuning of the hypothetical bounds of equation (4.3). The sparse reward function, however, can be transferred in a straight-forward way by adjusting the compensation reward component that is dependent on the maximum transportation time R_{comp}, the maximum processing time $T_{max\_processing}$, and the maximum number of operations $N_{operations}$ by using equation (5.3).

To prove that the results of the sparse reward were not positively influenced by the Q-Value masking, we repeated the experiments of the dense global reward design using Q-Value masking and observed whether the performance improved. However, the results with Q-Value masking became worse. This confirms that the comparison using Q-Value masking for the sparse reward design with the dense reward designs was competitive.

6.2.2   Cooperative Behavior

After we showed that agents improved the total makespan being trained on global rewards, it was crucial to analyze the actual agent behavior as it had to be proven that the improvement of the total makespan was achieved by cooperation and not by the improved individual agent performance (Pol et al., 2021). If this were the case, we would have had to apply further measures that encourage cooperation, such as schedules with cooperating agents that can lead to a better makespan than individual acting agents, as shown in figure 2.2. We therefore manually analyzed the Gantt charts after training with the dense local reward for 2,000 epochs and after training with the dense global reward for an additional 2,000 epochs. From this experiment we had a direct comparison whether the global reward affects the cooperative behavior. The analysis of the Gantt charts of the validation scenarios showed that there were cases where the total makespan decreased because of the improvement of the individual schedules of agents and not due to cooperation.

However, there were scenarios where the combination of job specifications did not require cooperation to achieve a near-optimal makespan. For other cases, cooperative behavior was identified where an agent selected a sub-optimal way to improve the performance of other agents (Pol, 2020). To prove the ability of cooperation, ten validation scenarios were created that explicitly required cooperative agent behavior, for instance where machines were needed by several agents in parallel. The analysis of the Gantt charts showed that agents cooperate where needed. Figure 6.3 shows one of these scenarios comparing the Gantt charts after 2,000 epochs using the dense local reward and after an additional 2,000 epochs of re-training using the dense global reward. Three agents (agent 1, agent 2, and agent 3) had to schedule three different jobs that were designed so that some operations collide. The first two operations of the first job collided with the first two operations of the second job. The second two operations of the second job collided with the second two operations of the third job (Pol et al., 2021).

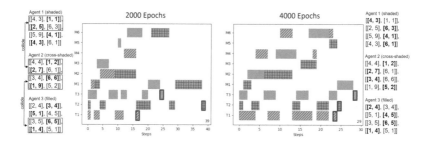

Figure 6.3 Cooperative Schedule comparison (Pol et al., 2021)

By comparing the two Gantt charts, it could be observed that the agents learned to improve the total makespan by choosing the worse option if it helps to avoid queues or other agents to be faster. Agent 1 and agent 2 initially chose $M1$ for their first operation and $M2$ for their second operation, which caused an unwanted queue in front of both of the machines. In the Gantt chart on the right, it can be observed that the agent 1 chose $M4$ and $M6$, which cost higher processing and transport times for the agent locally, but allowed agent 2 to choose $M1$ and $M2$ and to finish 15 time steps earlier (Pol et al., 2021). Agent 3 did not yet decide optimally, as it chose $M2$, thereby causing a queue and $M4$ instead of $M5$ for the second operation. Furthermore, agent 2 and agent 3 both took $M6$ and $M1$ for the third and fourth operations, respectively. After training with global rewards, agent 2 avoided the collision by choosing $M3$ and $M5$ instead (Pol et al., 2021). Even though the optimal schedule was not found in every case, it could be shown that cooperative behavior was achieved by the re-training using the global reward design.

6.3   Benchmarking against Heuristic Search Algorithms

After comparing the makespans of the different reward designs amongst each other, we evaluated the overall performance of the MARL approach by comparing it to classical search and optimization methods. Therefore, the algorithms "hill climb", "genetic algorithm", and "simulated annealing" were tested using the Python package mlrose (Hayes, 2019). Detailed information about the three algorithms are summarized in Russell Stuart and Norvig (2009). Our goal was to demonstrate that the performance of the trained RL agents used online during production can compete with the search algorithms that are used for calculating the schedule offline before production. We formulated the MrFJSP so that these methods could search for complete schedules in a given amount of time. The makespan of the schedules can afterwards be compared with the schedules of the MARL. It was also demonstrated how long it would take the conventional approaches to re-schedule during run-time. Simulated annealing outperformed hill climb and genetic algorithms and is therefore described and demonstrated in the following results (Pol et al., 2021). We designed the MrFJSP as a discrete-state optimization problem to apply mlrose using the problem class mlrose.DiscreteOpt. In this problem class, the state is defined as the overall result, which is the planned sequence of machines to be used for the jobs, and is also called the offline schedule. Considering three jobs and four operations per job, the state array has a length of 12. Integer values in the range of one to six represent the six available machines. The format of the state vector defines the schedule searched for the MrFJSP and is the output of the algorithm while there is no input vector. Considering only one fixed machine topology, there were 612 possible states. The optimization algorithm tries to find an offline schedule that minimizes the custom fitness function. We defined this fitness function (mlrose.CustomFitness) with a similar logic as the sparse global reward function. In general, the feedback was calculated from the resulting makespan of the schedule, including waiting times and transportation times, using a matrix that stores the distances between the machine interfaces. Changes in the machine topology are therefore considered by the fitness function. For invalid states, we penalized the algorithm, in a similar way to the dense local reward design, by adding a value to the makespan that was higher than any possible valid makespan. Invalid states do not provide resources to finish the three jobs. The maximum number of iterations and attempts were empirically defined by comparing the resulting makespans. A good parameter set led to a computation time of 15 minutes (Pol et al., 2021).

6.3.1   Evaluation for Unknown and Known Situations

As we already evaluated the different reward designs in section 6.2, the focus of figure 6.4 is on benchmarking the reward designs against simulated annealing. The ten unseen validation scenarios introduced in section 6.2 were used and the resulting makespans are represented by the box plot. The median and spread makespan values were the metrices used to evaluate the reward designs for the validation scenarios. The lower both values were, the better the learned scheduling algorithm with respect to the makespan. The median makespan of the local dense rewards trained for 2,000 epochs was 40 and the spread makespan values lay between 32 and 42, with one outlier of 61. If a single point was more than 17% relative away from the second maximum makespan, we displayed this value as a single dot and excluded it from the spread makespan metric. Simulated annealing had a lower median and lower spread makespan values than all other scheduling approaches. It performed 6% better with respect to the median than our best MARL approach using the sparse global reward for 4,000 epochs.

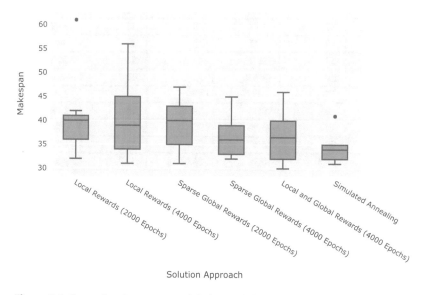

Figure 6.4 Comparison between reward designs and simulated annealing (Pol et al., 2021)

As the reward designs were only evaluated on unseen scenarios so far, we wanted to demonstrate how the best of the trained MARL algorithm works for known scenarios used in the training. The goal was to show the actual advantage of MARL in comparison to a search algorithm for known situations, as the trained NN equals a memory that performs very well and very fast for known states. Table 6.2 depicts the comparison of the sparse reward and simulated annealing using 20 scenarios, where the first 10 scenarios on the left side of the table were used in the training and the second 10 scenarios on the right side of the table are the unseen ones already used for the previous validation.

Table 6.2 Results of sparse rewards simulated annealing

MARL: training scenarios	SA: training scenarios		MARL: validation scenarios	SA: validation scenarios	
	Results	Improvement		Results	Improvement
41	40	2%	42	35	17%
34	27	21%	34	31	9%
32	37	−16%	38	34	11%
36	29	19%	33	32	3%
32	37	−16%	45	35	22%
28	30	−7%	32	41	−28%
37	37	0%	39	34	13%
29	29	0%	38	34	11%
38	42	−11%	32	32	0%
35	33	6%	33	35	−6%
34.2	34.1	0%	36.6	34.3	6%

We used equation (6.1) to demonstrate the relative improvement in percentage p between the results. Column three shows p between the MARL training after 4,000 epochs and the simulated annealing results for the training scenarios. Column six shows p respectively for the unseen validation scenarios. Comparing the average makespan of the sparse reward approach (34.2) to simulated annealing (34.1), there was no remarkable difference. However, there were some scheduling scenarios where simulated annealing performed worse, i.e., by 16% and others where simulated annealing performed better, i.e., by 21%. By comparing the makespans, we can state that the performance of the MARL approach can compete with simulated annealing for known training scenarios. For the unseen validation scenarios, simulated annealing performed 6% better on average, although there were two scenarios that performed worse than the MARL approach. Optimality was not achieved by either of the approaches. However, the overall goal of RQ2 was to meet the derived requirements with our solution, which does not include finding the absolute optimum. The derived requirements, for example, included the reaction to unforeseen situations (5), which is evaluated in the next section.

6.3.2 Evaluation of Real-time Decision-Making

It should be noted that simulated annealing achieved the results of the MARL system displayed in table 6.2 after around five minutes, and converged to the displayed results after 15 minutes (Pol et al., 2021). The greatest drawback of simulated annealing is that, even for the same scenario, it would take the same amount of time, as simulated annealing has no memory to store seen MrFJSPs. Simulated annealing could be enhanced by a memory, such as an NN, but it then has to be evaluated how the adopted algorithm performs for slightly different situations. The MARL system, in contrast, is designed for the appliance at runtime during production and returns the scheduling decisions in milliseconds, as only one forward pass of the trained NN is required per decision. Even for slightly different situations, the MARL solution provides meaningful results, as evaluated in the previous section. Therefore, our focus was less on optimality, but rather on the requirements for practical use, such as the reaction to unforeseen events (5) where real-time decision-making is needed. One example in Pol et al. (2021) demonstrates the advantage of the reactive scheduling system in comparison to the search algorithm. If a new manufacturing skill is added to a machine during run-time, e.g., by filling up the material buffer, the MARL system would recognize and consider this information through the updated job specification of the input state. Through the look-ahead, the agent can evaluate the new options of the subsequent operations and act accordingly. Using an offline scheduling approach, such as simulated annealing, the production must either wait for a complete re-schedule which can take minutes, or the new skill is neglected for the work-in-product jobs and is only considered for the next schedule calculation (Pol et al., 2021). We can summarize that our sparse reward design was competitive with the search algorithm in terms of the makespan and outperformed it for the requirement of real-time decision-making. It was therefore shown that the MARL solution meets requirement (5) by being able to react during run-time to any unforeseen event. The next section summarizes how well the developed solution meets the requirements by applying a consolidated and quantitative evaluation.

6.4   Consolidated Requirements Evaluation

The goal of this section is to consolidate the evaluation of the previous sections quantitatively as the last step of the evaluation process in figure 6.1. Thereby, we evaluated if the solution meets the derived requirements that were identified and ranked in the first step of the HoQ. Inspired by Gupta et al. (2017b), we applied the SIE model that was initially designed to evaluate the customer feedback on a product to identify whether the functionalities of a product meets the market's needs. Each requirement is mapped to a functionality that gets classified in the SIE diagram by the datum (satisfaction, importance). The diagram is divided into three areas that indicate whether the functionality was overly designed, frugally designed, or poorly designed. The frugal design area indicates that the satisfaction is as good as is needed for the relative importance. We adjusted the concept of Gupta et al. (2017b) and performed a self-assessment instead of a complete customer evaluation because it is accompanied by significant effort to find pilot customers that are willing to test and evaluate the solution. Table 6.3 presents the relevant functionalities which we implemented in our solution. The IDs represent the according requirements that are met by the functionalities.

The requirements were ranked in this order in chapter 2 and we evaluated the importance by numbering the requirements from 1, ..., 10, with 10 as the highest importance. To determine the satisfaction of these requirements, we defined three criteria that were evaluated in a self-assessment:

- Consideration in the concept: how well was the functionality considered in the concept?
- Sufficient evaluation: was the functionality evaluated generally enough or too specific?
- Evaluation results: how convincing were the evaluation results?

Inspired by Finkelstein et al. (2004), we performed a rating for each of the criteria with the rating scheme 1, ..., 10, with 10 indicating that a criterion is perfectly fulfilled. The average value of the three criteria determined the satisfaction of the requirement and is displayed in the column on the far right in the table of table 6.3.

Requirements (12) and (13) were well considered by the random sampling of the random generated job specifications. By the generic description of a product using the job specification in the advanced state encoding, new products can be handled directly. Requirement (15) concerning low engineering effort was considered throughout the design phase. The evaluation of (12) and (13) could have included more evaluation scenarios for a wider range of representative product variants. Nev-

Table 6.3 Assessment for the Satisfaction Importance Evaluation

ID	Functionality	Importance	Satisfaction			
			Concept	Evaluation	Results	Average
12	Random sampling + state encoding	10	10	7	9	9
13		9	10	8	10	9
16	Considering available machines in job specification	8	10	8	9	9
14	Considering available machines + random sampling	7	10	8	9	9
17	Decentral MARL + local rewards	6	6	5	5	5
18	Global reward design	5	10	8	8	9
6	Generalization	4	10	8	9	9
5	Situation-based decision-making	3	8	5	5	6
11	Petri net simulation	2	8	4	7	6
15	Self-learning MARL + Petri net	1	8	5	5	6

ertheless, the solution generalized very well for new products with arbitrary formats of job specifications. Requirements (16) and (14) concerned the consideration of overlapping machine skills and new machine skills with less effort. Both were considered by the job specification scheme that included all machines that are available and have the skills to perform a certain operation. The use of randomly generated job specifications did not limit the solution to the introduced FMS, but ensured that all kinds of jobs and machine set-ups could be handled automatically. One drawback could be the fixed number of machines that need to be defined for the input state, but this was tolerated so far as it was not a requirement.

Requirement (17) concerned the multi-objective optimization. The MARL approach itself enables independent agents to control products by their individual objectives. However, with the selected centralized learning approach, only one network is trained and instantiated n times for n agents. Different optimization objectives can be considered by training the network for different reward functions. This works as long as homogeneous agents are required that all strive towards the same goal. If different products should be manufactured by different optimization goals in parallel, the concept needs adjustments. The evaluation included experiments with local optimization goals and with local and global optimization goals, but only for the time-related makespans. Further experiments could be conducted using different optimization objectives such as energy consumption or quality. The satisfaction was thus rated moderately.

Requirement (18) was the global optimization goal and is one key aspect of the concept. It is fully considered by the global reward and proper state design. The sparse global reward design also fulfilled requirement (15), as less engineering effort is needed to re-design the sparse reward function for other FMS. The evaluation could have included more validation scenarios for a more representative result. The resulting makespans were good and could compete with heuristic search-based algorithms that were searching the solution for several minutes. In particular, the evaluation was mostly performed with unknown situations, which thereby fulfilled requirement (6). The solution generalized well, so that unknown situations could be handled without the need for re-training. The concept of using the MARL approach for online decision-making further fulfills requirement (5) of being able to react to unforeseen situations. The decisions were performed at each decision-making point based on the actual state input that included any changes that might come up during production. One forward pass of the trained NN takes milliseconds to perform the next decision, enabling the solution to be reactive to any event. No explicit evaluation was performed, as it was rather a requirement that was already considered in the concept design.

Any complex material flow (11) of an FMS can easily be modelled by the incidence matrix of the Petri net. However, it was not evaluated how the developed system performs with real complex material flows in bigger FMS. This remains an open topic for further research and the evaluation was thus rated lower. The presented results show agents that were able to learn a stable policy in the simulation and were thereby satisfying.

Figure 6.5 classifies the functionalities of the evaluated requirements in one of the three areas of the SIE model. Most of the functionalities met the satisfaction exactly as required for their level of importance and were classified in the frugal design. Functionalities (6) and (18) were slightly over designed. Nevertheless, the global optimization and the reaction to unknown situations are crucial aspects in the area of scheduling systems, even if they were rated as being of a lower importance by the customer rating in section 2.2.3. Therefore, it was legitimate to spend more effort than needed on these functionalities. The overall evaluation of the functionalities was positive and it can thus be concluded that the developed scheduling solution meets the requirements.

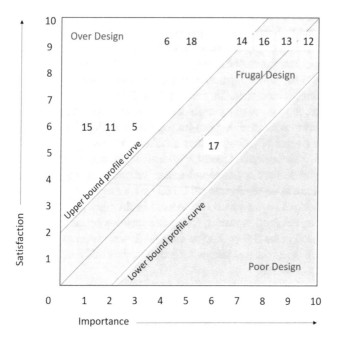

Figure 6.5 Satisfaction Importance Evaluation

6.5   Summary

After the design cycle that was performed in chapter 5, in this chapter we performed a quantitative evaluation of the developed solution. This evaluation aimed to answer RQ2 that had the goal of determining a suitable training concept, environment, and approach including reward designs to achieve reactive production scheduling meeting the derived requirements. We first evaluated the scalability of the solution and proved that it can handle a huge number of products and new products, including various formats, without re-training. Thereafter, we aimed to achieve the global objective through cooperating agents and therefore compared the dense and the sparse global reward design, which performed equally well. However, it should be noted that the engineering effort is smaller and the reproducibility is easier for the sparse reward design. The cooperative agent behavior was explicitly evaluated and proven using scenarios that require cooperation. We benchmarked the best MARL solution against a heuristic search algorithm based on the criteria of a minimized makespan and real-time decision-making. For the unknown validation scenarios, simulated annealing outperformed the sparse reward design by 6%. However, for known training scenarios, the average makespan of the sparse reward design could compete with the search algorithm. While simulated annealing performs well in creating offline schedules within minutes, in real-time scenarios with disturbances it would take too long to re-calculate the whole schedule. Our MARL solution, in contrast, performs reactive decision-making within milliseconds with one forward pass of the trained NN. We can thus conclude that our sparse reward design is competitive with the search algorithm in terms of the makespan and outperforms it when it comes to the reaction to unforeseen situations during run-time. In a self-assessment, we rated the functionalities that were mapped to the requirements in section 6.4 based on the criteria considered in the concept, as well as sufficient evaluation and evaluation results. Related to their importance, the functionalities were classified in the SIE evaluation diagram. It was observed that the developed solution well meets the requirements. From this, we could conclude that RQ2 was successfully answered.

Integration into a Flexible Manufacturing System

7

After having evaluated the developed scheduling solution in the previous chapter, in this chapter we focus on an integration concept of the solution into a real-world manufacturing system to answer RQ3. The concept and results presented in this chapter were published in Baer et al. (2020a).

The integration into existing FMSs is a practical problem and is thereby addressed by the design science research cycle. Design science is the study of new artifacts with the goal of solving real-world problems of general interest (Johannesson and Perjons, 2014). An artefact is an objective that is defined as the (partial) problem. The goal is to create, build, and evaluate new artifacts to improve the existing world. The artifact and knowledge about the artifact are outcomes for design science (Johannesson and Perjons, 2014). The design science research cycle describes the three pillars of building artifacts (Hevner, 2007). The relevance cycle involves the environment, which can include people and organizations that are interested in the artifact. The design cycle iterates between building and evaluating the artifact and the rigor cycle feeds back the derived knowledge to the scientific knowledge, i.e., through documentation, publications, or frameworks. In our case, the environment involves, e.g., the plant operator of the example FMS, who is interested in the integration of an online scheduling system in the existing FMS and provides relevant input for the artifact. We defined the artifact as the integration concept of our online scheduling solution in FMSs. We evaluated the artifact based on defined test scenarios for the example FMS which serves as a reference plant. The outcome should be a concept,

---

Supplementary Information The online version contains supplementary material available at https://doi.org/10.1007/978-3-658-39179-9_7.

S. Bär, *Generic Multi-Agent Reinforcement Learning Approach for Flexible Job-Shop Scheduling*, https://doi.org/10.1007/978-3-658-39179-9_7

as well as important findings and limitations that are published for all plant operators who are interested in integrating an online scheduling system in a brownfield FMS. The relevance cycle was applied to define the acceptance criteria in section 7.1 and the artifact was developed in section 7.2. The design cycle was applied to iteratively evaluate and improve the artifact in section 7.3. In section 7.4, we summarize the findings for the rigor cycle.

7.1   Acceptance Criteria for the Integration Concept

The acceptance criteria to address RQ3 were defined by the relevance cycle of the design science research cycle (Hevner, 2007). We therefore conducted a workshop together with the plant operator of the example FMS. We focused on acceptance criteria that determine the successful integration of the solution, as the MARL scheduling solution itself was already evaluated in chapter 6. The plant operator was asked how the design artifact can improve the environment and how the improvement can be measured. The improvements are summarized in table 7.1 in the column "acceptance criteria", and the measurements are summarized in the column "test scenarios". The evaluation methods for each acceptance criterion are provided.

Table 7.1 Acceptance criteria for artifact

ID	Acceptance criteria	Test scenarios	Evaluation Method
a	Functioning scheduling for various product variants	Scheduling ten orders with arbitrary product variants	Observational
b	Efficient production flow	Analysis of waiting times	Testing (white box)
c	Handling of unforeseen events	Take cart and place it to an arbitrary location at run-time	Experiment
d	Handling of new machine skills	New material is inserted into an arbitrary machine online	Experiment
e	Handling of new machines	New machine is docked to the system online	Experiment, interview, analysis

7.2 Integration Concept of MARL Scheduling Solution

In addition to the acceptance criteria of the plant operator, the derived requirements for the overall solution also had to be considered. For the integration concept, less engineering effort (15) is a relevant aspect that has to be taken into account. The integration concept in the MES and the information flow are described in the following section as the proposed artifact. The artifact was evaluated in the design cycle based on the acceptance criteria and was iteratively adjusted in section 7.3.

7.2.1 Integration in the MES

An MES can include several functionalities, such as data acquisition, order prioritization and order release, offline planning, and triggering the execution of the offline-plan (McClellan, 1997). Some MES can also include an online scheduling system or a hybrid approach with offline-planning that is updated frequently. In the proposed integration concept, the MES is enhanced by an online scheduling component. The scheduling system exchanges information with the MES and it therefore makes sense to either incorporate it into the MES or to provide the functionality as a service to the MES. As the concept should be re-usable, and not monolithic architecture, we decided for the service variant. Almada-Lobo (2015) anticipated in 2015 that MES will develop towards the approach that the MES software itself will be central with decentralized functionalities. In a similar way to what he described, our concept proposes that multiple agents control the products decentrally as one functionality of the MES. In the case where an MES already has an offline scheduling component, it can be enhanced by our reactive scheduling system in a complementary way. An MES with decentralized functionalities has the advantage of being flexible (Almada-Lobo, 2015). The offline scheduling system can be used before production starts for prioritizing and releasing the orders, while the reactive scheduling system controls the products during the production phase online. As a decentralized functionality of the MES, the MARL scheduling system does not interact with the FMS directly, but proposes the control actions for the MES. In the case where an MES does not include an offline scheduling system, our reactive scheduling system could be utilized to pre-plan the schedule offline and to determine the order release of the order stack.

The online scheduling service can either run on a cloud system or on-premise in the plant. With the cloud option, the solution has a high availability and can also be scaled automatically. As there are already on-premise server set-ups in the example FMS, the cloud variant would involved a higher effort in terms of costs, security, and maintenance. The MES runs on the on-premise server and it is reasonable to deploy the scheduling system on the same level for low latency. In other cases, where high availability and scalability are important, the service can be set up in the cloud.

The three relevant components for the production control are the MES, the FMS, and the Product Lifecycle Management (PLM) system. The RL agents interact with the MES and the agents do not control the FMS directly. The execution of the schedule is still performed by the MES, as it interacts with the field level, such as transport Programmable Logic Controller (PLC), by method calls that trigger specific functions in the PLC. The execution on the control level is performed by the PLC itself, i.e., by the transport PLC that controls the transport switches. This concept is chosen with the purpose of being able to integrate the agent-based scheduling system into an existing brownfield production control. Only the interface between the agent and the MES must be implemented to integrate the reactive scheduling solution. This has the advantage that the actual plant control does not need to be understood or changed.

The PLM system is a software system for the product development process that documents internal and external development processes of a product, including e.g., the connection to the Computer-Aided Design (CAD) and Enterprise Resource Planning (ERP) systems, for an effective communication between different groups related to the product development (Lee et al., 2008). Figure 7.1 gives an overview of the interaction between the three main components of our concept. The PLM system provides relevant data for the manufacturing process of the product to the MES that also receives information from the FMS. The MES processes the data, provides it to the scheduling component and executes the recommended actions towards the FMS.

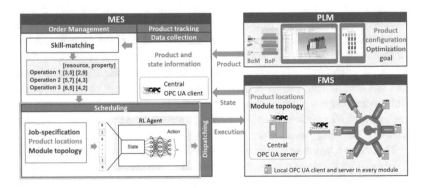

Figure 7.1 Interaction between components relevant for production control

The PLM system can have an interface to the product configuration system in which the product can be configured. The output of the configurator is the customer order that contains all product features. These product details are translated into the bill of processes and the bill of materials within the PLM system. This information serves as input for the MES, which uses the data for matching the capabilities of the available machines to the required processes. The requirement to perform this skill-matching is the capability-based semantic description of the resources within the FMS. This description, and how the skill matching is performed in detail, is not part of this work. Perzylo et al. (2019) introduced a concept to formally describe manufacturing skills to match them to the required manufacturing processes. Their concept is to enrich the existing information models, e.g., of a machine, by semantic annotations concerning their provided skills. The German Mechanical Engineering Industry Association (VDMA) develops Open Platform Communication Unified Architecture (OPC UA) (Burke et al., 2010) companion specifications for certain domains, such as robotics, to develop a standardized formalization of skills. This should enable a common understanding of skills, if they are used by the all machine builders within a domain (Perzylo et al., 2019). With these standardized descriptions of machine skills and manufacturing processes, the selection of resources can be performed. Resources and process steps are therefore compared based on the same skill types. If the properties of these skill types match, the resources are declared as being compatible. The DIN SPEC 92000 defines standardized semantics for the properties with the goal to enable an automatic skill-matching (Perzylo et al., 2019). The output of the skill-matching can have different formats. We required the MES to provide the job specification introduced in section 4.3. The job specification contains all available machines with their processing times per operation and is propagated to the MARL scheduling system that is highlighted by the box in figure 7.1. The state input for the agent instance requires the information about all product locations and the current machine topology in addition to the job specification. The execution of the recommended action is performed by the dispatching functionality of the MES. One option for integrating our solution into an FMS is using the industrial communication standard OPC UA for seamless communication between the MES and the field level. OPC UA servers and clients are utilized for the continuous exchange of information between the participants. Subscriptions can be used to perceive every relevant variable update and method call can be used to trigger functions in the PLC for the production control. The deployed communication between MES and the example FMS utilizing OPC UA is described in appendix B of the ESM.

7.2.2　Information Exchange

The sequence diagram in figure 7.2 demonstrates the proposed information flow and interaction between the instances of the RL agent, MES, transport PLC, and module PLC, RFID scanner, and transport switch.

When the scanner instance scans a cart, the cart ID is reported to the transport PLC (provide_location(cart_ID)). Together with the information of the scanner instance, the transport PLC knows where the controlled cart is located and requests the next action from the MES (request_action(cart_ID, location)). The cart can either be located on the transportation conveyor or on the conveyor in front of a machine. Based on this information, there are two alternatives that are carried out by the MES, as demonstrated by the divided sequence diagram and the conditions *location == conveyor* and *location == in front of machine*. If the cart is located on the transportation conveyor, the MES requests the agent of the MARL system for the next routing by providing the information about the cart_ID together with the location of the cart. If the MES only consisted of an offline scheduler, the MES would need to check if the cart_ID is aligned with the plan and re-plan on demand. Our concept extends the MES by an agent type, where each instance takes over the online control of one cart that is assigned to a product. The internal logic of the agent is not depicted and is discussed in section 4.2. The result of the agent is propagated back to the MES and passed through to the transport PLC that controls the relevant transport switch (control_routing (cart_ID): routing). Once the transport switch has executed the request, the feedback of a successful or unsuccessful execution is given back to the PLC and passed through to the MES. When the cart passes the scan point on

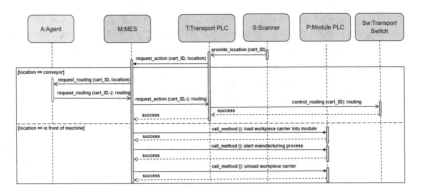

Figure 7.2 Sequence diagram of agent, MES, scanner, transport PLC, and switch

the conveyor in front of a machine, the MES directly triggers the method call to the module PLC to load the work-piece carrier into the machine, which means that it physically loads it into the working area of the machine. After a successful response from the module, the MES would trigger the method to start the process and to unload the work-piece carrier to the conveyor again after a successful processing.

7.3   Design Cycle

The design cycle is an iterative process between building and evaluation of the design artifact. Based on the acceptance criteria of section 7.1, the design was evaluated and every time a criterion could not be met, the design was adjusted. In the evaluation, all criteria need to be evaluated. In the following, we describe the addressed acceptance criteria, the evaluation method used, and the extent to which the design met the criteria.

7.3.1   Functioning Scheduling

The primary requirement of the plant operator is that various production orders can be scheduled online and get assigned to valid machines after the integration of our solution. To evaluate the applicability, we deployed the MARL scheduling solution to the MES of the example FMS as described in section 7.2.1. We observed and analyzed the output of the system, and more specifically whether the products were scheduled to valid machines at run-time. The acceptance criterion (a) of a functioning scheduling system was tested based on ten orders, each of which had three arbitrary product variants. The validation orders were not used in the training, which also validates the generalization aspect in the real-world FMS. Ten out of ten validation orders were successfully scheduled to valid machines after some minor bug-fixes on the PLC OPC UA interface. The first acceptance criterion was therefore met.

7.3.2   Efficient Production Flow

To further analyze the efficient production flow, we visually observed whether the production process of the ten validation orders from the previous test run efficiently as required by acceptance criterion (b). Even though the products were dispatched to the correct machines, we noticed that the production control did not run ideally as the carts with the products stopped in front of a switch on every scan point and waited for up to two seconds. It seemed as if the decision of the MES was not yet propagated to the field level once the cart arrived on the scanner of the relevant switch. As an evaluation method, we used white box testing where the input and output of every interface is logged and analyzed. As opposed to black box testing, we had the advantage of directly identifying problems in the information flow. We therefore analyzed whether the communication between the scanner, transport PLC, MES, and MARL system works in the proposed sequence of figure 7.2. The hypothesis was that the agents can be called when the cart arrives one decision-making point prior to the switch that should be controlled by this action. We logged the communication requests and responses of each component to identify the response times. We analyzed the log files and identified that the call sequence works as proposed in the initial concept, but with delays. A cart has to wait on each switch for one to two seconds. Therefore, we found that calling an agent to control the next switch after its cart has arrived at the decision-making point is too late. The time for receiving and processing the sensor data, providing the information to the agent, sending it back, and executing the decision causes the product to wait at the decision-making point for one to two seconds for the response. Acceptance criterion (b) was the efficient production flow of the products in the plant. As the waiting time constitutes approximate 20% of the transportation time between the decision-making point and the switch, the production flow was considered inefficient and would not be acceptable for manufacturing operators. We analyzed the following possible root causes for the delay:

- the latency, as the transmitting time of a message, can be hardly improved as the sender and receiver of the information had a small distance and there no intermediate devices, such as switches or repeaters, were used.
- mechanical delays, e.g., of the switch, are very small in relation to the waiting time of up to two seconds, and would therefore hardly improve the response time.
- the concept design might be sub-optimal. The design decision of calling the agent n decision-making points before the switch thus needed to be investigated.

As proposed in the design cycle, we adjusted the concept design. We therefore investigated the different alternatives for calling the agents that are opposed in table 7.2.

The alternative concepts call the agents on the decision-making points $n - 2$, $n - 3$, or $n - 4$ prior to the switch n. With a greater distance prior to the switch, the risk increases that the situation based on which the decision was taken is no longer

Table 7.2 Concept design of agent call logic

Concept	Risk
n-2	Other cart could overtake
n-3	Higher change to get overtaken
n-4	Situation has changed completely

up to date. The cart can for example be overtaken by another cart or machines can be occupied by other products. Therefore, we decided on concept $n-2$ as the preferred design. This adjustment has no effect on the decision-making of the agent, as it is merely a re-definition of the product location and the state input stays the same. For example, one agent is called immediately after the cart has left position 7 towards position 8, with the state input as if the product has already arrived at position 8. In reality, the product will arrive at a later point in time.

We again evaluated the conceptual adjustment of our design based on the criterion of an efficient production flow. The ten production orders were therefore scheduled using the new design. We observed the product behavior in the plant and noticed an improvement. The carts did not have to wait on the switch for the decision to be executed, but could pass directly after being stopped for a brief scan. After analyzing the log files of the communication requests and responses, no speed increase could be identified. This is because we did not improve the network response time but adjusted the design by using the concept of calling the agent one decision-making point earlier than before. The acceptance criterion (b) was successfully achieved, and the design adjustment was therefore used for further evaluation. Figure 7.3 shows the updated sequence diagram of the interaction of the scanner, transport PLC, MES, and Agents.

The main difference is that when the scanner provides the location of a cart, the transport PLC has already stored the destination for that cart in an internal variable table. The PLC can simply look up the destination and control the relevant transport switch directly. With this concept, the MES can asynchronously provide multiple transport commands for controlled products. Afterwards, the PLC provides the new cart location to the MES and receives the next destination for this cart that is again stored in the internal variable table. The remaining sequence stays unchanged as we simply re-defined the definition of the location of the cart. A particular position number now indicates that the cart is on the way to that location number, whereas in the initial concept the cart just left the position number. With this re-definition, and the introduction of the internal variable table, we ensured that there is sufficient time to process and exchange the information for the online production control.

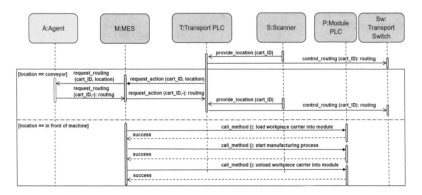

Figure 7.3 Updated sequence diagram of information exchange amongst the components

7.3.3 Handling of Unforeseen Events

Acceptance criterion (c) was the handling of unforeseen events. We already considered this for the development of the MARL solution through requirement (5) and evaluated the exception handling of the integration concept based on this criterion in an experiment. We selected an exception that occurred in the past and is relevant to the plant operator. The experimental set-up was defined by one order with one job, where the job specification is known to the observer of the experiment. To construct a conveyor breakdown as an unforeseen situation, the observer of the experiment removed the cart from the conveyor just before it wanted to enter the conveyor section in front of machine $M5$ to get processed there. The cart was then taken and placed on a different location on the main conveyor. The set-up of the experiment is depicted in figure 7.4. It was kept simple, with only two machines with different skills based on their available material. A job specification was selected that only needed the material of $M5$ so all operations had to be processed by that machine. The desired product is depicted on the right side and the bolt dot represents the controlled cart. It was manually taken from the conveyor after decision-making point twelve and was put back on the conveyor right before decision-making point nine.

The hypothesis was that the MARL system can handle the manual intervention of manually placing the cart in another conveyor section because the communication was set up so that every change of the relevant variables gets published to the MES automatically (details about the communication set-up are described in the appendix B. The concept of the MARL solution was designed in such a way that the agents perform their decision on every decision-making point independently from what

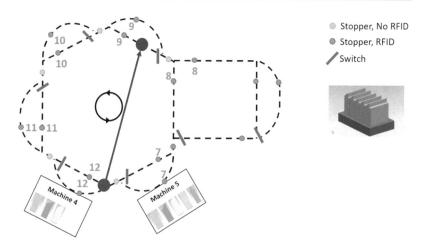

Figure 7.4 Re-placing the controlled product (bolt dot) during run-time

happened before, and solely based on the current state information (Baer et al., 2020a). We expected the agent of the product to perform the next decision-making based on the new state information of the FMS that was provided to the scheduling system without running into an error. The behavior of the cart at the decision-making points was observed during the experiment and the cart was controlled on the conveyor section again to $M5$ without any interruption. The experiment was successfully repeated five times with the introduced set-up. For all executed transportation requests, the PLC checked the destination of a cart against the desired destination that is stored in the internal data base, before the request was deleted from the list. In the case of a deviation, as in our experiments, the scenario was stored in a log file to be provided, i.e., to the maintenance personnel. The analysis of the log-files showed that for five out of five experiments, the agent received the updated information on the next decision-making point and could perform the action based on the new information. The hypothesis was therefore validated, and the experiment successfully proved the acceptance criterion (c).

7.3.4 Handling of New Machine Skills

Acceptance criterion (d) requires the integrated scheduling solution to be able to handle new machines' skills. In the example FMS, a new machine skill can either

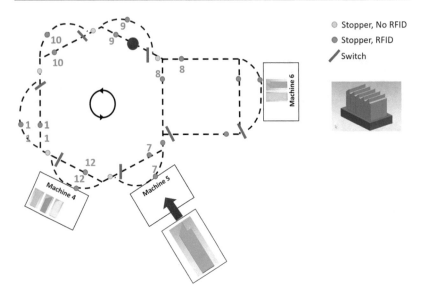

Figure 7.5 Filling up material of $M5$ during run-time

be realized by adding a new machine tool or material. We set up an experiment where we introduced a new skill to a machine by filling up the material of $M5$ during run-time. The hypothesis was that with the integration concept, the necessary information is provided to recognize and react on this during the run-time. Figure 7.5 depicts the scheme of the set-up in the example FMS. The desired product is shown on the right. We defined the set-up by three machines, where $M4$ had no material available that was needed by the job being produced and $M6$ had the available material. The required material was placed to $M5$ during run-time and was thereby enhanced with a new skill. We chose a set-up with only one job to ensure that there was no influence by other agents and that only the reaction on new skills could be tested. One job specification was used for all experimental runs. The experiment consisted of a control group that did not receive the experimental procedure, and an experimental group that received the experimental procedure. For the control group, ten runs were performed with $M4$, $M5$ (without material), and $M6$. The controlled cart was navigated to $M6$ to get the product assembled for all runs. For the experimental group, ten runs were conducted in which the material of $M5$ was filled up at arbitrary timings. The expectation was that the MARL system would navigate the controlled product to $M5$ that received the new skill, as it was closer and the makespan would be smaller. The observation was that for the control group, 10/10 runs were successfully completed and for the experimental group, 7/10 finished as expected by choosing $M5$. For 3/10, the cart was controlled to $M6$ instead of $M5$. To validate whether the behavior depends on the integration concept or on the agent's policy, we conducted the experiment in the simulation environment. The observation was that for the control group, 10/10 runs were finished successfully using $M6$ while in the experimental group, 10/10 runs were finished successfully using $M5$. As the behavior in the example plant deviates from this reference, we analyzed the information flow and the integration concept. As in the experimental group, the material was filled up at arbitrary timings, and the assumption was that the information was not yet received when the controlled cart passed $M5$. As suggested in the design cycle, we adjusted the integration concept and performed the experimental evaluation again. The material was filled up once the cart passed a defined decision-making point that was just far enough from the machine so that the information could be provided on when the cart arrived on the relevant decision-making point to select the machine. The experimental group was repeated considering this definition, and achieved 10/10 runs that were finished with the smaller makespan using $M5$. The hypothesis was thus validated, as the MARL solution is independent of new machine skills. The machine with the new skill is added to the specific operation in the job specification. The agent recognized the new machine skill by the updated job specification and dispatched the job to the machine accordingly. In our concept, the MES performs the skill-matching every time updated information about new machines or new machine skills is provided. The job specification is adjusted as a result, and the agent receives the new state information. To meet acceptance criterion (d), we defined a decision-making point the product has to pass when the material is filled up to receive the information in time. The hypothesis was therefore validated because, as with the integration concept, the MARL solution can handle updated machine skills.

7.3.5   Handling of New Machines

We finally tested the handling of a new machine that gets connected to the FMS during run-time to meet acceptance criterion (e). Similar to the approach in the previous section, we conducted an experiment with one control group using $M4$ and $M6$ and one experimental group where $M5$ was connected to one machine interface of the system during run-time. This set-up and the desired product are displayed in figure 7.6. The product can be assembled using $M4$ and $M6$. Once $M5$ is available, it can be assembled using $M5$ only because all the necessary materials are available. In the experimental group, the machine was connected once the cart passes the same decision-making point that was used to fill up the material in the previous section. The hypothesis was that the MARL prefers $M5$ once it is available in the experimental group. For the evaluation of the experiment, we observed the behavior of the controlled cart. For 10/10 runs of the control group, the cart was navigated to $M4$ and afterwards to $M6$. The unexpected result was that 0/10 runs of the experimental group were successful because a run-time error occurred. The cart was controlled to the conveyor section in front of the connected $M5$. In a conversation with the plant operator, we learned that the MES sent the method call to load the cart into the machine when the error occurred. We interviewed the plant operator about the docking and booting process of a machine and the analysis of the interview led to the following sequence of actions for starting a machine that is docked to the FMS:

1. Selection of the machine instance and docking position on Human Machine Interface HMI (e.g., "Assembly machine five on docking position five")
2. Triggering the docking process on HMI
3. Manufacturing module is locked pneumatically by the locking system
4. 24 V supply voltage gets released at the docking position by the backbone

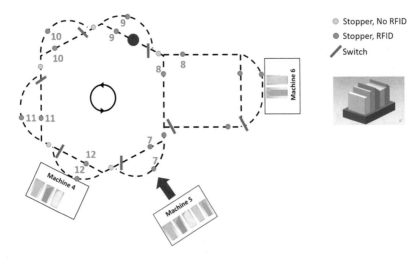

Figure 7.6 Docking a new machine to the system during run-time

5. PLC and peripheral systems (e.g. HMI) boot
6. HMI Run-time triggers slightly to supply the security switch with energy
7. The security switch sends an Dynamic Host Configuration Protocol (DHCP) request to the backbone DHCP server
8. **The manufacturing module hands over a self-description of available skills to a knowledge base**
9. The backbone server triggers safety PLC to activate the safety program for the manufacturing module at a defined docking position
10. The input and outputs signal of the safety ET200SP of the manufacturing module get monitored by the safety PLC of the backbone
11. A safe state of the manufacturing module leads to the release of the operation voltage (400 V Alternating Current (AC))

We determined that the machine was still in the booting process and not yet idle when the MES tried to call the load skill. The crucial point within this sequence is that the self-description of the available skills is handed over in step eight, before the safety modules of the machine were started and triggered the 400 V AC power supply. In the design cycle, we arrived at the solution of shifting action number eight towards the end as action number twelve. Our analyses revealed that a similar problem can happen when undocking a machine, because the sequence for the undocking process is only reverted, meaning that the agent could dispatch the controlled product to the machine even if it is not available anymore. Adjusting the order of the action sequence when docking a machine to the FMS is an important finding for the integration of the MARL system. The implementation was not straightforward in the case of the example FMS, and we therefore implemented a workaround for repeating the experiment. We measured the time used for steps nine to eleven and set it as waiting time before the self-description is handed over in step eight. Steps nine to eleven are asynchronous and can be performed during this waiting time. With the scheduling solution that was used previously in example FMS, this issue did not arise as the time for re-calculating a new schedule took longer than the booting process of the machine. Using the adjusted action sequence, we repeated the runs of the experimental group. We connected the machine to the system before production start to ensure that the machine had finished booting when the cart was on the defined decision-making point. We then observed that in 10/10 runs the controlled cart was successfully dispatched to $M5$. We validated the hypothesis of being able to react to new machines with the integration concept and thereby met the last acceptance criterion (e).

7.4   Summary

We presented the integration concept of the MARL scheduling system into the OT system of an existing FMS to answer RQ3. The design science research cycle was applied to build and develop an integration concept. Five acceptance criteria were defined with the operator of the example FMS in the relevance cycle. The design was evaluated in the design cycle by means of these acceptance criteria, and the rigor cycle feeds back the artifact and findings to the scientific community. The first criterion (a) was to have a functioning online scheduling for various products, which was successfully tested based on ten orders with three arbitrary jobs. We further analyzed the ten production cycles to evaluate whether the production flow was efficient (b). Unwanted waiting times on the stoppers in front of the switches were observed and the concept of the sequence diagram was adjusted to the logic depicted in figure 7.3. We re-defined to call the agents two decision-making points prior to the switch instead of only one prior to the switch, but with the same state information, as it was one decision-making point prior to the switch. To evaluate the handling of unforeseen events, which was defined under acceptance criterion (c), we constructed a conveyor issue by re-positioning a cart to another decision-making point. As the MARL system performs situation-based decision-making, the agent used the new state information and controlled the cart to the desired machine without interruption. Acceptance criterion (d) was fulfilled by the successful experiment of inserting a new machine skill during the run-time. When a machine is filled up with new material after the cart passed a defined decision-making point, the information was provided to the agent in time. In a similar approach as that used for evaluating the response to the new machine skills, we evaluated the connection of a machine during run-time (e). After the adjustment of the action sequence, the information was provided to the MARL system in time and was successfully considered in the decision-making. Using the iteration between the building and evaluating of the design cycle, we identified aspects that needed to be adjusted to meet the acceptance criteria. The clear contribution of this research to the rigor cycle is published in (Baer et al., 2020a) and can be summarized as follows:

- Integration of the MARL system in the MES as a service
- Definition of decision-making points, so that the information is available in time
- Notification of availability of a machine only after booting has completed

The concept for integrating the MARL scheduling system into an existing manufacturing system was presented and evaluated. We successfully addressed and answered RQ3, as the design artifact was evaluated as being efficient and solves the problem it was designed for.

Critical Discussion and Outlook

8

Chapters 5 to 7 presented an approach of a reactive scheduling solution for FMS. This chapter now reflects the presented approach and discusses opportunities for further research, thereby performing step ten of the HoQ. In discussion rounds with researchers and experts, we analyzed the results to identify aspects that were not yet considered in the concept because they were not listed in the requirements, but which could become requirements in the future. We therefore highlight aspects of being independent of the maximum number of machines and jobs and consider heterogeneous optimization objectives. We also found that instead of conducting parameter studies, learning hyper-parameters could be helpful in terms of lower engineering effort, and that the guidelines for trustworthy AI can become relevant during product development and deployment of the solution in productive FMS.

In the results in section 6.1, we demonstrated that the solution is flexible in terms of the number of operations and the number of alternatives per operation. Therefore, one major component is the state design of the MARL solution that enables various job specification formats. However, we defined the size of the state vector by the maximum number of machines, queue positions, and number of agents, which prevents the scaling of the problem, i.e., for a larger number of machines, without re-training the agents for the new problem size. The solution could be improved by making it independent of these components. A possible approach to circumvent the dependency on the number of machines would be to navigate the carts to machine interfaces instead of the machine, which has the advantage that the training can be performed with a fixed number of machine interfaces while the number of docked machines can change. The filtering of the available skills to execute an operation and the mapping to the machine interface could be the preparatory step to generate the job specification. In this way, the information on the machine topology in the state becomes obsolete. To further scale the solution to the number of jobs, respectively agents, eligibility criteria could be implemented to

S. Bär, *Generic Multi-Agent Reinforcement Learning Approach for Flexible Job-Shop Scheduling*, https://doi.org/10.1007/978-3-658-39179-9_8

select the most relevant agents for the state vector, e.g., by a central agent. There are alternatives to incorporate the selection of relevant agents in the agents' tasks, as in the learning of coordination and coordinating Q-Learning proposed by Wiering and van Otterlo (2012).

In the evaluation of the MARL solution, we considered jobs that need to be fulfilled by the minimum makespan. Our concept can only be followed as long as homogeneous agents with the same optimization objective are used, as we trained only one network that can be used by different agents as independent sub-policies. The advantage is that we thereby created a stable environment despite the MARL set-up, because the policy of the network changes in the same direction for all agents. For considering individual optimization objectives, separate networks need to be trained with individual reward functions. Additional learning methods, such as concurrent learning, therefore need to be investigated (Panait and Luke, 2005).

In section 5.2, we demonstrated the process of using parameter studies to determine hyper-parameters for the reward function based on the respective FMS. These hyper-parameters are for example dependent on the size of the FMS and on the transportation time on the conveyors. In future, it could be investigated whether these hyper-parameters could be learned based on meta-information of the FMS to further reduce the engineering effort for setting up the MARL scheduling solution.

We evaluated the dense reward and the sparse reward design, which led to similar scheduling performance. The sparse reward design is dependent on the number of operations of a job used in the training, as the resulting maximum reward that is propagated back after finishing all jobs is related to the total makespan. The assumption is that the trained agents can schedule jobs of different lengths equally well, as the state space only consists of an extract of the job specification. To validate this hypothesis, future work needs to be carried out to evaluate how well the MARL solution trained with a sparse reward can handle job specifications of various sizes and lengths. Further investigations could also assess whether the learning behavior using the sparse reward function decays with scheduling tasks of growing size. With longer planning horizons in the training, the credit assignment problem increases, which makes it more difficult for the agent to learn a stable behavior.

We successfully observed and validated cooperative agent behavior in section 6.2.2. With selected job specifications that required cooperative behavior we proved that agents consider the demands of other agents for machines and choose an alternative if possible. However, there can be cases where two agents require the same machine and both choose to select the alternative to grant the other agent access to the preferred machine. In such cases, it would be counterproductive if the decisions are not coordinated properly. In our evaluation, such behavior could not be observed, but it remains for future work to conduct deeper investigations into this aspect. Communicating the intentions between agents could help to train them to coordinate their cooperating actions.

The evaluation of the MARL solution was performed based on criteria that were raised by interviews with interested stakeholders and the integration concept was evaluated based on criteria defined with the operator of the example FMS. Future work could focus on the evaluation of the concept according to the guidelines of trustworthy AI. The High-Level Expert Group on AI presented these guidelines to the European Commission in April 2019. The following seven key requirements should be met by AI solutions to be considered as trustworthy: human agency and oversight, technical robustness and safety, privacy and data governance, transparency, diversity, non-discrimination and fairness, environmental and societal well-being, and accountability.

In the evaluation of the integration concept in chapter 7, we had to adjust the location of the decision-making point to control the next routing of the cart because the information was not yet received when the cart arrived at the switch. The location of the decision-making point is $n - x$, where n is the location of the switch. Thereafter, x was defined to be two for the example FMS used to keep the distance to the controlled switch as small as possible. We thereby decreased the risk that the state of the FMS has changed completely by the time the cart arrives at the switch and that the decision was made based on older information that is no longer applicable. The location $n - 2$ for the decision-making point was tested in the example FMS, and we proved that the decision of the agent had already arrived when the controlled cart reached the switch. This design was empirically determined and cannot be generalized to other FMSs. It remains for future research to either find a general concept for the definition of the distance between the decision-making point and controlled resource or to iteratively learn the best design. Further, we can consider extending the integration concept in such a way that the PLC prepares transportation control in advance before the controlled cart arrives at the scanner at the switch. This has the advantage that the cart must no longer stop and be scanned at the switch, thereby improving the speed and efficiency.

Finally, further evaluation of the MARL solution has to be carried out to assess the stability of the system when it is used over a longer period of time in the example FMS. With the evaluation in the long-run, the solution can be tested on scenarios that might not have been considered in the training and that need to be specifically re-trained in specific situations. For exceptional handling cases in the FMS, log files with the state and actions of the agents are saved. Future work could utilize these log files for a selective re-training of agents, e.g., in order to improve cooperation of the agents in complex situations. This is an aspect that would need to be implemented in future work to continuously improve the MARL scheduling concept during run-time.

Summary

Manufacturers have to align their organizations to stay flexible when facing changing market demands. Reactive scheduling is one crucial component of production control that should be flexible, e.g., in terms of available machines, new product variants, and unforeseen events. With the contribution of a reactive scheduling system and the integration concept for existing FMSs, we successfully achieved the determined goals defined by the three research questions. With the problem formalization as MrFJSP and the concept of MARL agents controlling carts, we answered RQ1 in chapter 4. The training approach, including the selection of suitable algorithms to achieve a global optimization addressed RQ2 in chapter 5. The integration concept for existing FMS addressed RQ3 in chapter 7.

We utilized the HoQ to map the market demands to the technical implementation. In the first step, we identified prioritized requirements from expert interviews to define the extended scheduling problem for FMS in chapter 2 and successfully evaluated our concept based on these requirements in chapter 4 to answer RQ1. We developed a MARL scheduling system with agents controlling carts that are assigned to jobs. The transport could thereby be controlled in addition to the assignments to machines on dedicated decision-making points. In section 6.2.2, we proved that agents are proactive and cooperate with each other and in section 7.3.3, it was shown that agents are reactive by being able to change their route on each decision-making point. One major aspect of the work was the development of a suitable reward design. In the dense reward design, a local reward was given depending on the resulting processing, waiting, or transport time of the controlled cart. The local reward was subsequently adjusted by the global reward factor, which was calculated at the end of an episode to motivate the agents to cooperate. The sparse reward design, in contrast, was determined once after an episode, and was inversely related to the total makespan. The results showed that both reward functions performed equally well for the validation scenarios used. However, the engineering effort for the sparse

S. Bär, *Generic Multi-Agent Reinforcement Learning Approach for Flexible Job-Shop Scheduling*, https://doi.org/10.1007/978-3-658-39179-9_9

reward is lower, as it can be calculated by only using the total makespan. Through the successful evaluation of the concept, we answered RQ1.

The training strategy, including the selection of suitable algorithms to achieve a common optimization objective addresses RQ2 in chapter 5. The proposed solution was developed iteratively by parameter studies and evaluation cycles of possible designs. In section 5.3, we demonstrated that a training strategy utilizing a centralized learning approach with one network and a collective replay memory has the advantage of creating a stable environment despite the MARL set-up. Different agents use the trained network as independent sub-policies during run-time. The best approach to present scenarios to the RL agents is to randomly sample triples of jobs from a pre-defined training set of job specifications every five epochs, as shown in section 5.4. We successfully evaluated the generalization to unknown situations representing new product variants with arbitrary job specification formats and new machine skills in chapter 6. In the benchmarking with Simulated Annealing, the average makespan achieved and the reaction time on unforeseen situations were compared in section 6.3.1. We found that the achieved average makespan of simulated annealing was 6% higher than the results of MARL solution. However, the proposed reactive scheduling concept outperforms simulated annealing with respect to the reaction time on unforeseen situations during run-time. We successfully evaluated the MARL scheduling system based on the derived requirements summarized in the satisfaction importance evaluation in section 6.4 and thereby answered RQ2.

The integration concept for existing FMSs in chapter 7 addresses RQ3. The MARL approach enhanced existing MES by the service of reactive scheduling during run-time. The communication to the field level for information exchange and execution of the production control remains the task of the MES. Using the design science research cycle, the definition of the decision-making point prior to the controlled switch was re-defined, so that the decision is there when the controlled cart arrives at the scanner of the switch. The reaction to new machine skills, new machines, and failed conveyor sections was successfully shown.

The critical discussion in chapter 8 addresses the opportunities for further work. Future research could connect in scaling the concept to a larger amount of jobs controlled in parallel, while carts could be controlled to machine interfaces instead of machines being independent of the machine topology. A long-term evaluation could finally assess stability and utilize log-files of scenarios with sub-optimal decisions to re-train agents for continuous improvement.

Bibliography

Joseph Adams, Egon Balas, and Daniel Zawack. The shifting bottleneck procedure for job shop scheduling. *Management Science*, 34(3): 91–401, 1988. ISSN 00251909, 15265501. URL http://www.jstor.org/stable/2632051.

Francisco Almada-Lobo. The industry 4.0 revolution and the future of manufacturing execution systems (mes). *Journal of innovation management*, 3(4): 16–21, 2015.

Marcin Andrychowicz, Filip Wolski, Alex Ray, Jonas Schneider, Rachel Fong, Peter Welinder, Bob McGrew, Josh Tobin, Pieter Abbeel, and Wojciech Zaremba. Hindsight experience replay, 2017. URL http://arxiv.org/pdf/1707.01495v3.

Schirin Baer, Jupiter Bakakeu, Richard Meyes, and Tobias Meisen. Multi-agent reinforcement learning for job shop scheduling in flexible manufacturing systems. In *2019 Second International Conference on Artificial Intelligence for Industries (AI4I)*. IEEE, 2019.

Schirin Baer, Felix Baer, Danielle Turner, Sebastian Pol, and Tobias Meisen. Integration of a reactive scheduling solution using reinforcement learning in a manfacturing system. In *Automation 2020*, Bade-Baden, Germany, 2020a.VDI.

Schirin Baer, Danielle Turner, Punit Kumar Mohanty, Vladimir Samsonov, Romuald Jupiter Bakakeu, and Tobias Meisen. Multi agent deep q-network approach for online job shop scheduling in flexible manufacturing. In *ICMSMM 2020: International Conference on Manufacturing System and Multiple Machines*, Tokyo, Japan, 2020b.

Schirin Baer, Sebastian Pol, Danielle Turner, Felix Baer, and Tobias Meisen. Scaling a reinforcement learning approach for online job shop scheduling in flexible manufacturing systems to various products. In *Automation 2021, VDI-Bericht*, pages 53 – 60, Baden-Baden, Germany, 2021. VDI.

R. Bellman. Dynamic programming. *Princeton univ. press." NJ 95*, 1957.

Irwan Bello, Hieu Pham, Quoc V. Le, Mohammad Norouzi, and Samy Bengio. Neural combinatorial optimization with reinforcement learning, 2017.

Dimitri P. Bertsekas and John N. Tsitsiklis. *Neuro-dynamic programming*, volume 3 of *Athena scientific optimization and computation series*. Athena Scientific, Belmont, Mass., 2. print edition, ca. 1999. ISBN 978-1-886529-10-6.

S Binato, WJ Hery, DM Loewenstern, and MAURICIO GC Resende. A grasp for job shop scheduling. In *Essays and surveys in metaheuristics*, pages 59–79. Springer, 2002.

Daan Bloembergen, Karl Tuyls, Daniel Hennes, and Michael Kaisers. Evolutionary dynamics of multi-agent learning: A survey. *Journal of Artificial Intelligence Research*, 53: 659–697, 2015. 10.1613/jair.4818.

© The Editor(s) (if applicable) and The Author(s), under exclusive license to
Springer Fachmedien Wiesbaden GmbH, part of Springer Nature 2022
S. Bär, *Generic Multi-Agent Reinforcement Learning Approach for Flexible Job-Shop Scheduling*, https://doi.org/10.1007/978-3-658-39179-9

Léon Bottou. Large-scale machine learning with stochastic gradient descent. In Physica-Verlag HD, editor, *Proceedings of COMPSTAT'2010*, pages 177–186, 2010.

Paolo Brandimarte. Routing and scheduling in a flexible job shop by tabu search. *Annals of Operations research*, 41(3):157–183, 1993.

Peter Brucker, Bernd Jurisch, and Bernd Sievers. A branch and bound algorithm for the job-shop scheduling problem. *Discrete applied mathematics*, 49(1–3):107–127, 1994.

Thomas J Burke, Jürgen Lange, and Frank Iwanitz. Opc-from data access to unified architecture. *Published: VDE VERLAG GMBH*, 2010.

Lucian Busoniu, Robert Babuska, and Bart de Schutter. A comprehensive survey of multiagent reinforcement learning. *IEEE Transactions on Systems, Man, and Cybernetics, Part C (Applications and Reviews)*, 38(2):156–172, 2008. ISSN 1094-6977. 10.1109/TSMCC.2007.913919.

MWPV Supply Chain. A supply chain consultant evalutaion of kiva systems (amazonrobotics), 2021. URL https://mwpvl.com/html/kiva_systems.html.

Felix T. S. Chan and Jie Zhang. Modelling for agile manufacturing systems. *International Journal of Production Research*, 39(11):2313–2332, 2001. 10.1080/00207540110051932. URL https://doi.org/10.1080/00207540110051932.

George Chryssolouris. *Manufacturing systems: theory and practice*. Springer Science & Business Media, 2013.

Caroline Claus and Craig Boutilier. The dynamics of reinforcement learning in cooperative multiagent systems. *AAAI/IAAI*, 1998(746–752):2, 1998.

B. C. Csáji and L. Monostori. Csáji, balázs csanád, and lászló monostori. adaptive stochastic resource control: a machine learning approach. In *Journal of Artificial Intelligence Research*, pages 453–486. 2008.

Balázs Csanád Csáji and László Monostori. Adaptive algorithms in distributed resource allocation. In *Proceedings of the 6th international workshop on emergent synthesis (IWES 06)*. Citeseer, 2006.

S Esquivel, S Ferrero, R Gallard, Carolina Salto, Hugo Alfonso, and Martin Schütz. Enhanced evolutionary algorithms for single and multiobjective optimization in the job shop scheduling problem. *Knowledge-Based Systems*, 15(1–2):13–25, 2002.

Jerzy Filar and Koos Vrieze. *Competitive Markov Decision Processes*. Springer New York, New York, NY, 1997. ISBN 978-1-4612-4054-9. 10.1007/978-1-4612-4054-9. URL http://dx.doi.org/10.1007/978-1-4612-4054-9.

Anthony Finkelstein, Clare Gryce, and Joe Lewis-Bowen. Relating requirements and architectures: A study of data-grids. *Journal of Grid Computing*, 2(3):207–222, 2004.

Krzysztof Foit, Grzegorz Gołda, and Adrian Kampa. Integration and evaluation of intralogistics processes in flexible production systems based on oee metrics, with the use of computer modelling and simulation of agvs. *Processes*, 8(12):1648, 2020.

Thomas Gabel. *Multi-Agent Reinforcement Learning Approaches for Distributed Job-Shop Scheduling Problems*. 2009.

Thomas Gabel and Martin Riedmiller. Adaptive reactive job-shop scheduling with reinforcement learning agents. In Citeseer, editor, *International Journal of Information Technology and Intelligent Computing*, pages 14–18. 2008.

Michael R Garey, David S Johnson, and Ravi Sethi. The complexity of flowshop and jobshop scheduling. *Mathematics of operations research*, 1(2):117–129, 1976.

Chris Gaskett, Luke Fletcher, and Alexander Zelinsky. Reinforcement learning for a vision based mobile robot. In *Proceedings. 2000 IEEE/RSJ International Conference on Intelligent Robots and Systems (IROS 2000)(Cat. No. 00CH37113)*, volume 1, pages 403–409. IEEE, 2000.

Lagoudakis M. Parr R. Guestrin, C. Coordinated reinforcement learning. In *Proceedings of the 19th International Conference on Machine Learning*, pages pp. 227–234, 2002a.

Venkataraman S. & Koller D. Guestrin, C. Context-specific multiagent coordination and planning with factored mdps. In *In AAAI/IAAI*, pages pp. 253–259, 2002b.

Jayesh K. Gupta, Maxim Egorov, and Mykel Kochenderfer. Cooperative multi-agent control using deep reinforcement learning. In Gita Sukthankar and Juan A. Rodriguez-Aguilar, editors, *Autonomous Agents and Multiagent Systems*, volume 10642 of *Lecture Notes in Computer Science*, pages 66–83. Springer International Publishing, Cham, 2017a. ISBN 978-3-319-71681-7. https://doi.org/10.1007/978-3-319-71682-4_5.

Ravi Kumar Gupta, Farouk Belkadi, Alain Bernard, et al. Evaluation and management of customer feedback to include market dynamics into product development: Satisfaction importance evaluation (sie) model. In *DS 87-4 Proceedings of the 21st International Conference on Engineering Design (ICED 17) Vol 4: Design Methods and Tools, Vancouver, Canada, 21-25.08. 2017*, pages 327–336, 2017b.

Jean Harb and Doina Precup. Investigating recurrence and eligibility traces in deep q-networks, 2017. URL http://arxiv.org/pdf/1704.05495v1.

Peter E Hart, Nils J Nilsson, and Bertram Raphael. A formal basis for the heuristic determination of minimum cost paths. *IEEE transactions on Systems Science and Cybernetics*, 4(2):100–107, 1968.

NAJ Hastings and C-H Yeh. Job oriented production scheduling. *European Journal of Operational Research*, 47(1):35–48, 1990.

G Hayes. mlrose: Machine Learning, Randomized Optimization and Search package for Python, 2019. URL https://github.com/gkhayes/mlrose.

Z He, Taeyong Yang, and DE Deal. A multiple-pass heuristic rule for job shop scheduling with due dates. *THE INTERNATIONAL JOURNAL OF PRODUCTION RESEARCH*, 31(11):2677–2692, 1993.

Alan R Hevner. A three cycle view of design science research. *Scandinavian journal of information systems*, 19(2):4, 2007.

Hong-Zhong Huang, Yanfeng Li, Wenhai Liu, Yu Liu, and Zhonglai Wang. Evaluation and decision of products conceptual design schemes based on customer requirements. *Journal of mechanical science and technology*, 25(9):2413–2425, 2011.

Walter Huber. *Produktion der Zukunft*, pages 261–264. Springer Fachmedien Wiesbaden, Wiesbaden, 2016. ISBN 978-3-658-12732-9. https://doi.org/10.1007/978-3-658-12732-9_9. URL https://doi.org/10.1007/978-3-658-12732-9_9.

Johann Hurink, Bernd Jurisch, and Monika Thole. Tabu search for the job-shop scheduling problem with multi-purpose machines. *Operations-Research-Spektrum*, 15(4):205–215, 1994.

Chris Meyer Ian MacKenzie and Steve Noble. How retailers can keep up with consumers, 2013. URL https://www.mckinsey.com/industries/retail/our-insights/how-retailers-can-keep-up-with-consumers#.

Paul Johannesson and Erik Perjons. *An introduction to design science*. Springer, 2014.

Michał Kempka, Marek Wydmuch, Grzegorz Runc, Jakub Toczek, and Wojciech Jaśkowski. Vizdoom: A doom-based ai research platform for visual reinforcement learning. In *2016 IEEE Conference on Computational Intelligence and Games (CIG)*, pages 1–8. IEEE, 2016.

Barry Kirwan. Soft systems, hard lessons. *Applied Ergonomics*, 31(6):663 – 678, 2000. ISSN 0003-6870. https://doi.org/10.1016/S0003-6870(00)00041-7. URL http://www.sciencedirect.com/science/article/pii/S0003687000000417. Fundamental Reviews in Applied Ergonomics.

't Hoen P. Bakker B. Vlassis N. Kok, J. Utile coordination: Learning interdependencies among cooperative agents. In *Proceedings of the IEEE Symposium on Computational Intelligence and Games (CIG 2005)*, pages 29–36, 2005.

Vlassis N. Kok, J. Sparse cooperative q-learning. In *Proceedings of the 21st International Conference on Machine Learning*, 2004.

Vlassis N. Kok, J. Collaborative multiagent reinforcement learning by payoff propagation. In *Journal of Machine Learning Research*, pages 1789–1828. 2006.

Wouter Kool, Herke van Hoof, and Max Welling. Attention, learn to solve routing problems!, 2019.

Yoram Koren, Uwe Heisel, Francesco Jovane, Toshimichi Moriwaki, Gumter Pritschow, Galip Ulsoy, and Hendrik Van Brussel. Reconfigurable manufacturing systems. *CIRP annals*, 48(2):527–540, 1999.

Andreas Kuhnle, Jan-Philipp Kaiser, Felix Theiß, Nicole Stricker, and Gisela Lanza. Designing an adaptive production control system using reinforcement learning. *Journal of Intelligent Manufacturing*, pages 1–22, 2020.

Yu-Cheng Lee, Liang-Chyau Sheu, and Yuan-Gan Tsou. Quality function deployment implementation based on fuzzy kano model: An application in plm system. *Computers & Industrial Engineering*, 55(1): 48–63, 2008. ISSN 0360-8352. https://doi.org/10.1016/j.cie.2007.11.014. URL https://www.sciencedirect.com/science/article/pii/S0360835207002719.

Victor R Lesser. Cooperative multiagent systems: A personal view of the state of the art. *IEEE Transactions on knowledge and data engineering*, 11(1):133–142, 1999.

Timothy P Lillicrap, Jonathan J Hunt, Alexander Pritzel, Nicolas Heess, Tom Erez, Yuval Tassa, David Silver, and Daan Wierstra. Continuous control with deep reinforcement learning. *arXiv preprint* arXiv:1509.02971, 2015.

Lin Lin, Xin-Chang Hao, Mitsuo Gen, and Jung-Bok Jo. Network modeling and evolutionary optimization for scheduling in manufacturing. *Journal of Intelligent Manufacturing*, 23(6):2237–2253, 2012.

Long-Ji Lin. *Reinforcement learning for robots using neural networks: Technical Report*. DTIC Document, 1993.

Ryan Lowe, Yi Wu, Aviv Tamar, Jean Harb, Pieter Abbeel, and Igor Mordatch. Multi-agent actor-critic for mixed cooperative-competitive environments, 2017. URL http://arxiv.org/pdf/1706.02275v3.

M. Montemerlo M. Riedmiller and H. Dahlkamp. Learning to drive a real car in 20 minutes. In IEEE Computer Society, editor, *Frontiers in the Convergence of Bioscience and Information Technologies*, pages 645–650. IEEE Computer Society, 2007.

Alessandro Mascis and Dario Pacciarelli. Job-shop scheduling with blocking and no-wait constraints. *European Journal of Operational Research*, 143(3):498–517, 2002.

Egorov Maxim. Multi-agent deep reinforcement learning, 2016.

Michael McClellan. *Applying manufacturing execution systems*. CRC Press, 1997.

Toni Melfi. Smart factory, 2015.

Volodymyr Mnih, Koray Kavukcuoglu, David Silver, Alex Graves, Ioannis Antonoglou, Daan Wierstra, and Martin Riedmiller. Playing atari with deep reinforcement learning, 2013. URL http://arxiv.org/pdf/1312.5602v1.

Volodymyr Mnih, Koray Kavukcuoglu, David Silver, Andrei A. Rusu, Joel Veness, Marc G. Bellemare, Alex Graves, Martin Riedmiller, Andreas K. Fidjeland, Georg Ostrovski, Stig Petersen, Charles Beattie, Amir Sadik, Ioannis Antonoglou, Helen King, Dharshan Kumaran, Daan Wierstra, Shane Legg, and Demis Hassabis. Human-level control through deep reinforcement learning. *Nature*, 518(7540):529–533, 2015. https://doi.org/10.1038/nature14236.

Volodymyr Mnih, Adrià Puigdomènech Badia, Mehdi Mirza, Alex Graves, Timothy P. Lillicrap, Tim Harley, David Silver, and Koray Kavukcuoglu. Asynchronous methods for deep reinforcement learning. In *International Conference on Machine Learning 2016*, pages 1928–1937, 2016. URL http://arxiv.org/pdf/1602.01783v2.

Punit Mohanty. *Multi-Agent Reinforcement Learning for Reactive Job Shop Scheduling in Flexible Manufacturing Systems*. 2020.

László Monostori, József Váncza, and Soundar RT Kumara. Agent-based systems for manufacturing. *CIRP annals*, 55(2):697–720, 2006.

Eugeniusz Nowicki and Czeslaw Smutnicki. A fast taboo search algorithm for the job shop problem. *Management science*, 42(6):797–813, 1996.

Peng Si Ow and Thomas E Morton. Filtered beam search in scheduling. *The International Journal Of Production Research*, 26(1):35–62, 1988.

Liviu Panait and Sean Luke. Cooperative multi-agent learning: The state of the art. *Autonomous Agents and Multi-Agent Systems*, 11(3):387–434, 2005. ISSN 1387-2532. https://doi.org/10.1007/s10458-005-2631-2.

Shrikant S Panwalkar and Wafik Iskander. A survey of scheduling rules. *Operations research*, 25(1):45–61, 1977.

Alexander Perzylo, Julian Grothoff, Levi Lucio, Michael Weser, Somayeh Malakuti, Pierre Venet, Vincent Aravantinos, and Torben Deppe. Capability-based semantic interoperability of manufacturing resources: A basys 4.0 perspective. *IFAC-PapersOnLine*, 52(13):1590–1596, 2019.

Michael Pinedo and Khosrow Hadavi. Scheduling: theory, algorithms and systems development. In *Operations Research Proceedings 1991*, pages 35–42. Springer, 1992.

Sebastian Pol. Cooperative behavior in a multi-agent reinforcement learning system for reactive scheduling in flexible manufacturing, 2020.

Sebastian Pol, Schirin Baer, Danielle Turner, Vladimir Samsonov, and Tobias Meisen. Global reward design for cooperative agents to achieve flexible production control under real-time constraints. In *Proceedings of 23rd International Conference on Enterprise Information Systems (ICEIS)*, Virtual, 2021. SCITEPRESS.

Dev Bahadur Poudel. Coordinating hundreds of cooperative, autonomous robots in a warehouse. *Jan*, 27:1–13, 2013.

Shuhui Qu, Jie Wang, Shivani Govil, and James O Leckie. Optimized adaptive scheduling of a manufacturing process system with multi-skill workforce and multiple machine types: An ontology-based, multi-agent reinforcement learning approach. *Procedia CIRP*, 57:55–60, 2016.

Michael Rabl. *Quality Function Deployment*, pages 127–142. Gabler, Wiesbaden, 2009. ISBN 978-3-8349-8780-8. https://doi.org/10.1007/978-3-8349-8780-8_10. URL https://doi.org/10.1007/978-3-8349-8780-8_10.

Amazon Robotics. Amazon robotics, 2015. URL https://www.amazonrobotics.com/#/.

Martin Roesch, Christian Linder, Christian Bruckdorfer, Andrea Hohmann, and Gunther Reinhart. Industrial load management using multi-agent reinforcement learning for rescheduling. In *2019 Second International Conference on Artificial Intelligence for Industries (AI4I)*, pages 99–102. IEEE, 2019.

Bernard Roy and B Sussmann. Les problemes d'ordonnancement avec contraintes disjonctives. *Note ds*, 9, 1964.

Sebastian Ruder. An overview of gradient descent optimization algorithms, 2016. URL http://arxiv.org/pdf/1609.04747v2.

J Russell Stuart and Peter Norvig. *Artificial intelligence: a modern approach*. Prentice Hall 2009.

Vladimir Samsonov, Marco Kemmerling, Maren Paegert, Daniel Lütticke, Frederick Christian Sauermann, Andreas Gützlaff, Günther Schuh, and Tobias Meisen. Manufacturing control in job shop environments with reinforcement learning. In *Proceedings of the 13th International Conference on Agents and Artificial Intelligence. Volume 1: Online, 04-06.02.2021*, pages 589–597, [Sétubal], 2021. 13th International Conference on Agents and Artificial Intelligence, online, 4 Feb 2021 - 6 Feb 2021, SCITEPRESS - Science and Technology Publications, Lda. https://doi.org/10.5220/0010202405890597. URL https://publications.rwth-aachen.de/record/814806.

Robert Schmitt, Christian Brecher, Burkhard Corves, Thomas Gries, Sabina Jeschke, Fritz Klocke, Peter Loosen, Walter Michaeli, Rainer Müller, Reinhard Poprawe, et al. Self-optimising production systems. In *Integrative Production Technology for High-Wage Countries*, pages 697–986. Springer, 2012. ISBN 978-3-642-21067-9. https://doi.org/10.1007/978-3-642-21067-9_6. URL https://doi.org/10.1007/978-3-642-21067-9_6.

Johannes MJ Schutten. Practical job shop scheduling. *Annals of Operations Research*, 83:161–178, 1998.

G Seliger and D Kruetzfeldt. Agent-based approach for assembly control. *CIRP Annals*, 48(1):21–24, 1999.

Abdul Munaf Shaik, VVS Kesava Rao, and Ch Srinivasa Rao. Development of modular manufacturing systems–a review. *The International Journal of Advanced Manufacturing Technology*, 76(5-8):789–802, 2015.

Satinder P. Singh and Richard S. Sutton. Reinforcement learning with replacing eligibility traces. In *Machine learning*, pages 123–158. 1996.

Mehdi Souier, Mohammed Dahane, and Fouad Maliki. An nsga-ii-based multiobjective approach for real-time routing selection in a flexible manufacturing system under uncertainty and reliability constraints. *The International Journal of Advanced Manufacturing Technology*, 100(9-12):2813–2829, 2019.

Peter Stone and Manuela Veloso. *Autonomous Robots*, 8(3):345–383, 2000. ISSN 09295593. https://doi.org/10.1023/A:1008942012299.

Richard S. Sutton and Andrew G. Barto. *Reinforcement learning: An introduction*. A Bradford book. MIT Press, Cambridge, Mass., [nachdr.] edition, 2010. ISBN 0-262-19398-1.

Ardi Tampuu, Tambet Matiisen, Dorian Kodelja, Ilya Kuzovkin, Kristjan Korjus, Juhan Aru, Jaan Aru, and Raul Vicente. Multiagent cooperation and competition with deep reinforcement learning. *PloS one*, 12(4):e0172395, 2017. https://doi.org/10.1371/journal.pone.0172395.

Ming Tan. Multi-agent reinforcement learning: Independent vs. cooperative agents. In Morgan Kaufmann Publishers Inc, editor, *Proceedings of the Tenth International Conference on Machine Learning*, pages 330–337, 1993. ISBN 1-55860-307-7.

Gerald Tesauro. Td-gammon, a self-teaching backgammon program, achieves master-level play. *Neural Computation*, 6(2):215–219, 1994. ISSN 0899-7667. https://doi.org/10.1162/neco.1994.6.2.215.

Frank Thomas, Christoph Kilger, Ralf Hermann, and Ralf Baginski. *Informationstechnologie als Treiber der Intralogistik*, pages 193–237. Springer Berlin Heidelberg, Berlin, Heidelberg, 2006. ISBN 978-3-540-29658-4. https://doi.org/10.1007/978-3-540-29658-4_5. URL https://doi.org/10.1007/978-3-540-29658-4_5.

Eiji Uchibe and Kenji Doya. Competitive-cooperative-concurrent reinforcement learning with importance sampling. In *Proc. of International Conference on Simulation of Adaptive Behavior: From Animals and Animats*, pages 287–296, 2004.

Umar Ali Umar, MKA Ariffin, N Ismail, and SH Tang. Hybrid multiobjective genetic algorithms for integrated dynamic scheduling and routing of jobs and automated-guided vehicle (agv) in flexible manufacturing systems (fms) environment. *The International Journal of Advanced Manufacturing Technology*, 81(9-12):2123–2141, 2015.

Robert Johannes Maria Vaessens, Emile HL Aarts, and Jan Karel Lenstra. Job shop scheduling by local search. *INFORMS Journal on computing*, 8(3):302–317, 1996.

Gohar Vahdati, Mehdi Yaghoubi, Mahdieh Poostchi, et al. A new approach to solve traveling salesman problem using genetic algorithm based on heuristic crossover and mutation operator. In *2009 International Conference of Soft Computing and Pattern Recognition*, pages 112–116. IEEE, 2009.

Hendrik Van Brussel, Jo Wyns, Paul Valckenaers, Luc Bongaerts, and Patrick Peeters. Reference architecture for holonic manufacturing systems: Prosa. *Computers in industry*, 37(3):255–274, 1998.

Peter JM Van Laarhoven, Emile HL Aarts, and Jan Karel Lenstra. Job shop scheduling by simulated annealing. *Operations research*, 40(1):113–125, 1992.

De Hauwere Y.M. Nowe A. Vrancx, P. Transfer learning for multi-agent coordination. In *Proceedings of the 3th International Conference on Agents and Artificial Intelligence*, pages pp. 263–272, 2011.

Bernd Waschneck, André Reichstaller, Lenz Belzner, Thomas Altenmüller, Thomas Bauernhansl, Alexander Knapp, and Andreas Kyek. Deep reinforcement learning for semiconductor production scheduling. In *2018 29th annual SEMI advanced semiconductor manufacturing conference (ASMC)*, pages 301–306. IEEE, 2018.

Christopher J. C. H. Watkins and Peter Dayan. Q-learning. *Machine Learning*, 8(3-4):279–292, 1992. ISSN 0885-6125. https://doi.org/10.1007/BF00992698.

Christopher John Cornish Hellaby Watkins. *Learning from Delayed Rewards*. PhD thesis, King's College, Cambridge, UK, May 1989. URL http://www.cs.rhul.ac.uk/~chrisw/new_thesis.pdf.

Gerhard Weiss. *Multiagent systems*. Intelligent robotics and autonomous agents. MIT Press, Cambridge, Mass., 2. ed. edition, 2013. ISBN 9780262018890. URL http://search.ebscohost.com/login.aspx?direct=true&scope=site&db=nlebk&db=nlabk&AN=550664.

Marco Wiering and Martijn van Otterlo. *Reinforcement Learning*, volume 12. Springer Berlin Heidelberg, Berlin, Heidelberg, 2012. ISBN 978-3-642-27644-6. https://doi.org/10.1007/978-3-642-27645-3.

J Wisner. Survey of smaller machine shops presents a useful profile by which to benchmark your own shop's operations. *Modern Machine Shop*, 66:86–86, 1993.

Teresa Wu, Nong Ye, and Dawei Zhang. Comparison of distributed methods for resource allocation. *International Journal of Production Research*, 43(3):515–536, 2005.

Peter R Wurman, Raffaello D'Andrea, and Mick Mountz. Coordinating hundreds of cooperative, autonomous vehicles in warehouses. *AI magazine*, 29(1):9–9, 2008.

P. Vrancx Y-M. De Hauwere and A. Nowé. Learning multi-agent state space representations. In ifaamas, editor, *Proc. of 9th Int. Conf. on Autonomous Agents and Multiagent Systems (AAMAS 2010)*, pages 715–722, 2010.

A Nowé YM De Hauwere, P Vrancx. Detecting and solving future multi-agent interactions. In *Proceedings of the AAMAS Workshop on Adaptive and Learning Agents*, pages 45–52, 2011.

Paraskevi T Zacharia and Elias K Xidias. Agv routing and motion planning in a flexible manufacturing system using a fuzzy-based genetic algorithm. *The International Journal of Advanced Manufacturing Technology*, 109(7):1801–1813, 2020.

Jian Zhang, Guofu Ding, Yisheng Zou, Shengfeng Qin, and Jianlin Fu. Review of job shop scheduling research and its new perspectives under industry 4.0. *Journal of Intelligent Manufacturing*, 30(4):1809–1830, 2019.

Wei Zhang and Thomas G Dietterich. A reinforcement learning approach to job-shop scheduling. In *IJCAI*, volume 95, pages 1114–1120. Citeseer, 1995.

Jing-ying Zhao. Evaluation index system of production planning in manufacturing enterprise. In Chen-Fu Chien, Ershi Qi, and Runliang Dou, editors, *IE&EM 2019*, pages 261–268, Singapore, 2020. Springer Singapore. ISBN 978-981-15-4530-6.

MengChu Zhou and Kurapati Venkatesh. *Modeling, simulation, and control of flexible manufacturing systems: a Petri net approach*. World Scientific, 1999.

Printed in the United States
by Baker & Taylor Publisher Services